Mosquitoes Suck!

Katherine Richardson Bruna, Sara Erickson, and Lyric Bartholomay

Comic by Bob Hall and Judy Diamond

Art by Bob Hall, Michael Cavallaro, Bob Camp, and Mike Edholm

Posters and production design by Aaron Sutherlen

Mosquitoes and Me
Urban Ecosystem Project

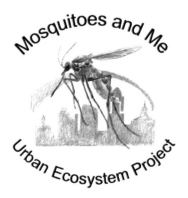

This project is funded by the National Institutes of Health #R25GM129210 (2015-2020). The content is solely the responsibility of the authors and does not necessarily represent the official views of the NIH.

http://www.research.hs.iastate.edu/urban-ecosystem-project/

UNIVERSITY OF NEBRASKA PRESS
LINCOLN

Manufactured in the United States of America. ♾

Library of Congress Cataloging-in-Publication Data

Names: Richardson Bruna, Katherine, author. | Hall, Bob, 1944 October 16– illustrator.

Title: Mosquitoes suck! / Katherine Richardson Bruna, Sara Erickson, Lyric Bartholomay; comic by Bob Hall and Judy Diamond; art by Bob Hall, Michael Cavallaro, Bob Camp, Mike Edholm; posters and production design by Aaron Sutherlen.

Description: Lincoln: University of Nebraska Press, [2021] | Mosquitoes and me urban ecosystem project. This project is funded by the National Institutes of Health #R25GM129210 (2015–2020).

Identifiers: LCCN 2020055770 | ISBN 9781496224347 (paperback)

Subjects: LCSH: Mosquitoes—Comic books, strips, etc.—Juvenile literature.

Classification: LCC QL536 .R495 2021 | DDC 595.77/2—dc23

LC record available at https://lccn.loc.gov/2020055770

CONTENTS

GRAPHIC STORIES

ESSAYS

POSTERS

AUTHORS' NOTE

Mosquitoes SUCK! celebrates our relationships with the talented young scientists of the Mosquitoes & Me summer camps in Des Moines, Iowa. Working with them, we created learning experiences driven by curiosity about this tiny but tenacious insect. Parts of our summer camp now come to life in the vibrant science comic you are holding in your hand. We especially want to recognize the young scientists of our Comic Book Club family. The mosquito comics they created showed us how far we can go with our big brains, an understanding of science, and some imagination. They also helped us bring the characters of *Mosquitoes SUCK!* to life.

None of our work would have been possible without generous funding from the National Institutes of Health Science Education Partnership Award.

After reading this science comic, we hope you'll agree that, while mosquitoes may suck, they're also pretty cool.

Mosquito scientists unite!

IOWA STATE UNIVERSITY
OF SCIENCE AND TECHNOLOGY

SEPA SCIENCE EDUCATION PARTNERSHIP AWARD
SUPPORTED BY THE NATIONAL INSTITUTES OF HEALTH

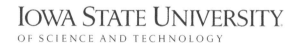

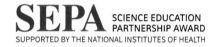

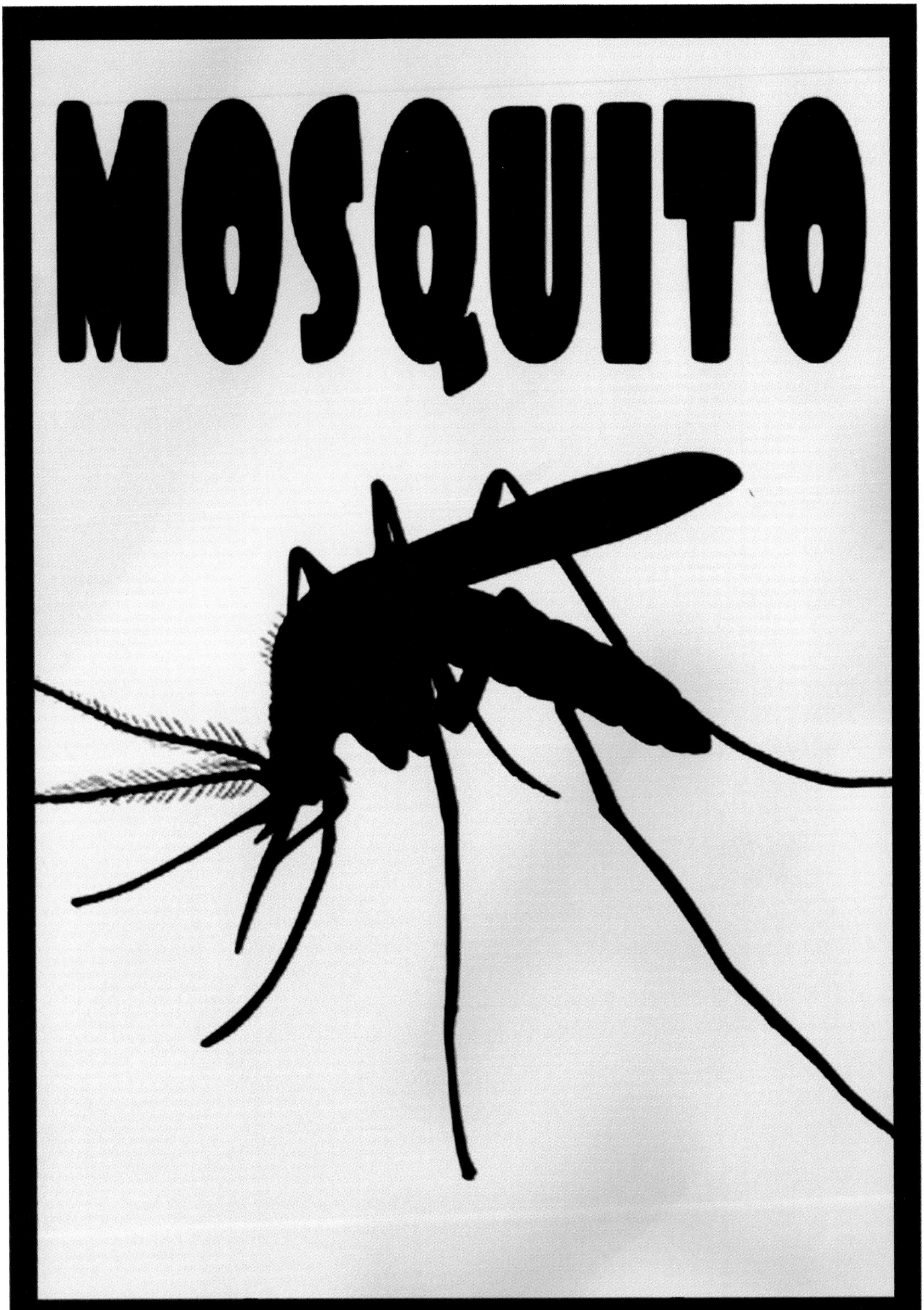
MOSQUITO

MOSQUITO MUSEUM

Story by Bob Hall and Judy Diamond

Illustrated by Bob Hall

Lettering by Mike Edholm

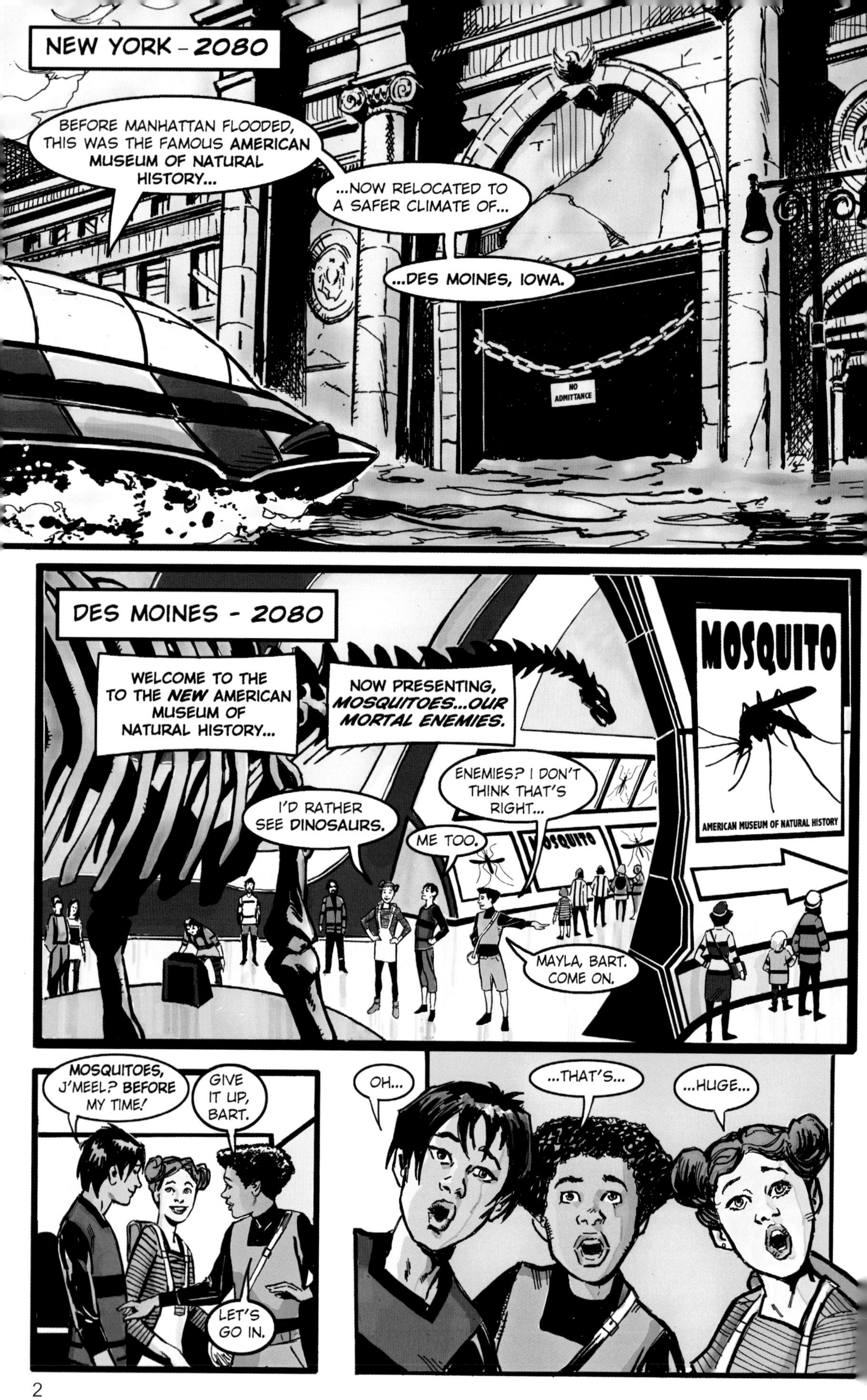

NEW YORK - 2080
BEFORE MANHATTAN FLOODED, THIS WAS THE FAMOUS **AMERICAN MUSEUM OF NATURAL HISTORY...**
...NOW RELOCATED TO A SAFER CLIMATE OF...
...DES MOINES, IOWA.
NO ADMITTANCE
DES MOINES - 2080
WELCOME TO THE TO THE *NEW* AMERICAN MUSEUM OF NATURAL HISTORY...
NOW PRESENTING, ***MOSQUITOES...OUR MORTAL ENEMIES.***
MOSQUITO
AMERICAN MUSEUM OF NATURAL HISTORY
I'D RATHER SEE **DINOSAURS.**
ME TOO.
ENEMIES? I DON'T THINK THAT'S RIGHT...
MOSQUITO
MAYLA, BART. COME ON.
MOSQUITOES, J'MEEL? **BEFORE** MY TIME!
GIVE IT UP, BART.
LET'S GO IN.
OH...
...THAT'S...
...HUGE...

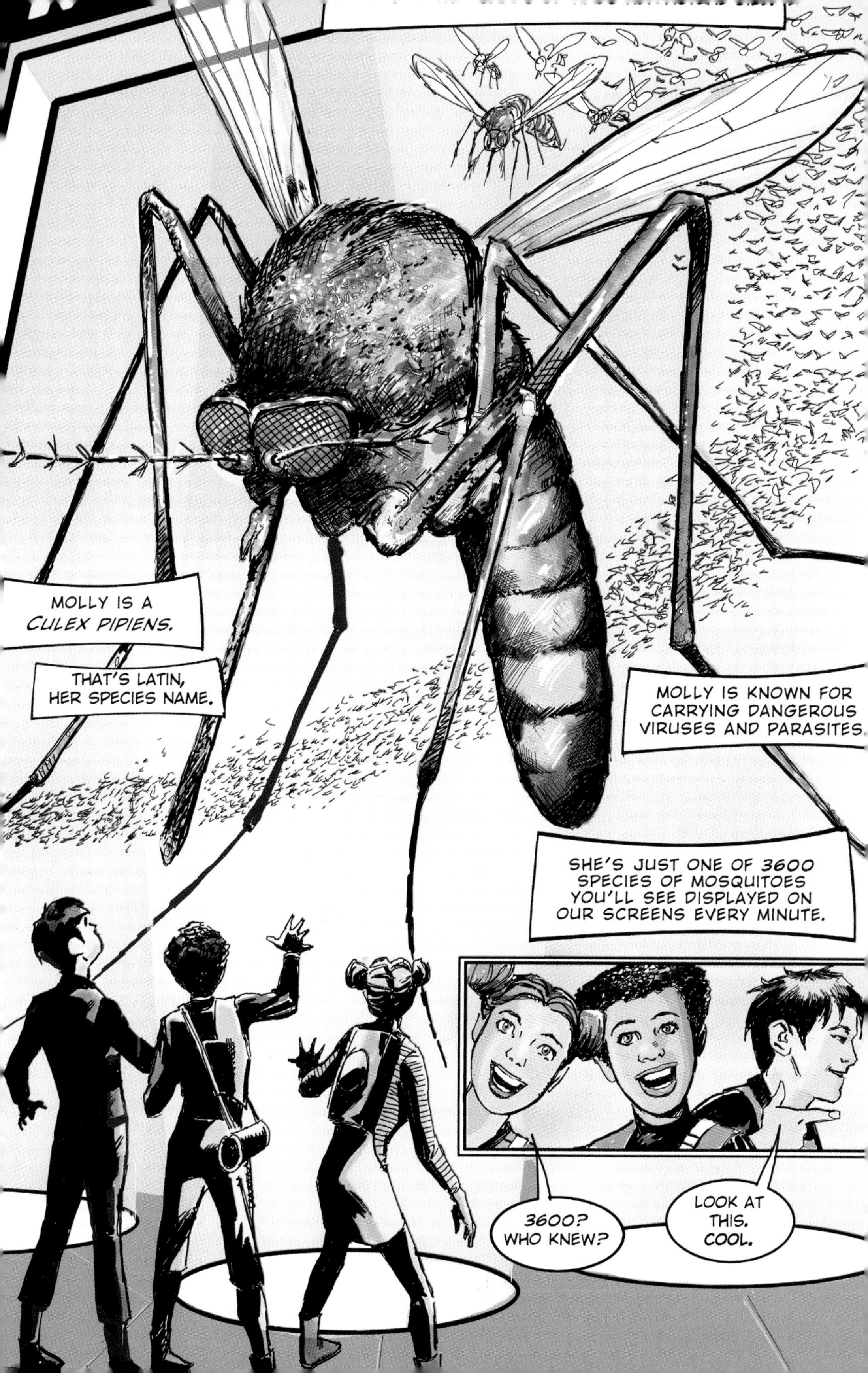
MOLLY IS A *CULEX PIPIENS.*
THAT'S LATIN, HER SPECIES NAME.
MOLLY IS KNOWN FOR CARRYING DANGEROUS VIRUSES AND PARASITES.
SHE'S JUST ONE OF 3600 SPECIES OF MOSQUITOES YOU'LL SEE DISPLAYED ON OUR SCREENS EVERY MINUTE.
3600? WHO KNEW?
LOOK AT THIS. COOL.

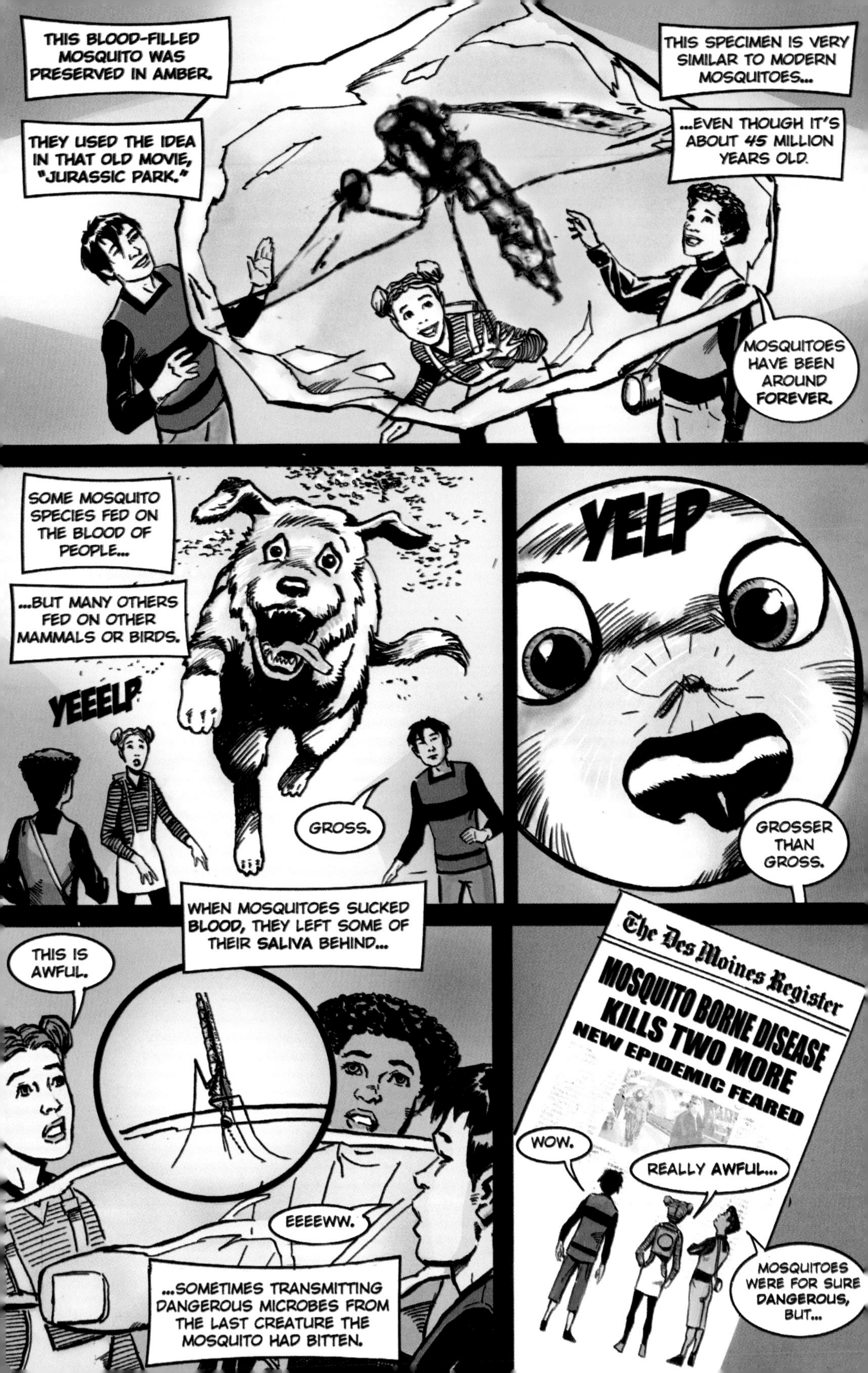
THIS BLOOD-FILLED MOSQUITO WAS PRESERVED IN AMBER.
THEY USED THE IDEA IN THAT OLD MOVIE, "JURASSIC PARK."
THIS SPECIMEN IS VERY SIMILAR TO MODERN MOSQUITOES...
...EVEN THOUGH IT'S ABOUT 45 MILLION YEARS OLD.
MOSQUITOES HAVE BEEN AROUND FOREVER.
SOME MOSQUITO SPECIES FED ON THE BLOOD OF PEOPLE...
...BUT MANY OTHERS FED ON OTHER MAMMALS OR BIRDS.
YEEELP
GROSS.
YELP
GROSSER THAN GROSS.
THIS IS AWFUL.
WHEN MOSQUITOES SUCKED BLOOD, THEY LEFT SOME OF THEIR SALIVA BEHIND...
EEEEWW.
...SOMETIMES TRANSMITTING DANGEROUS MICROBES FROM THE LAST CREATURE THE MOSQUITO HAD BITTEN.
The Des Moines Register
MOSQUITO BORNE DISEASE KILLS TWO MORE
NEW EPIDEMIC FEARED
WOW.
REALLY AWFUL...
MOSQUITOES WERE FOR SURE DANGEROUS, BUT...

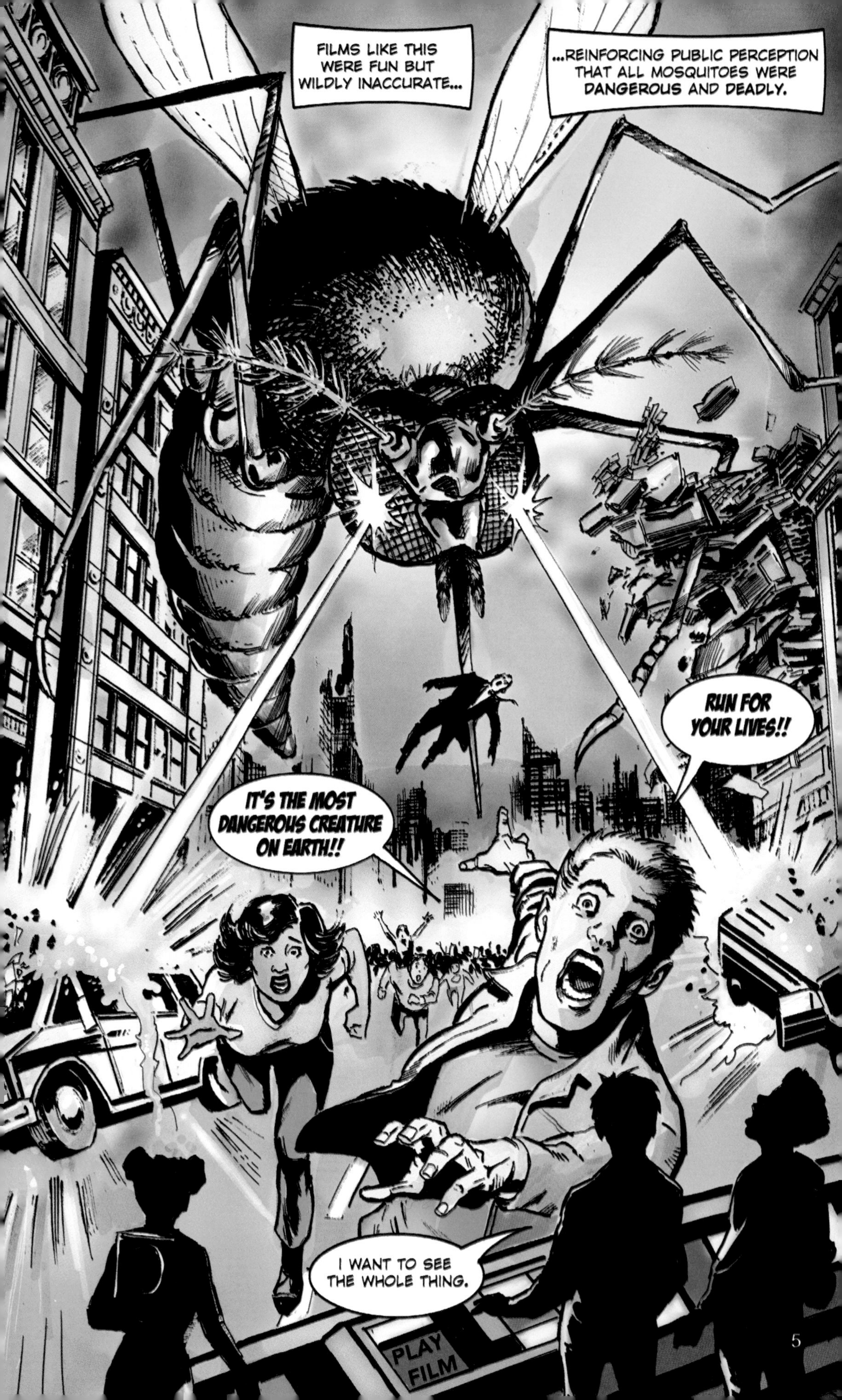
FILMS LIKE THIS WERE FUN BUT WILDLY INACCURATE...
...REINFORCING PUBLIC PERCEPTION THAT ALL MOSQUITOES WERE DANGEROUS AND DEADLY.
RUN FOR YOUR LIVES!!
IT'S THE MOST DANGEROUS CREATURE ON EARTH!!
I WANT TO SEE THE WHOLE THING.
PLAY FILM

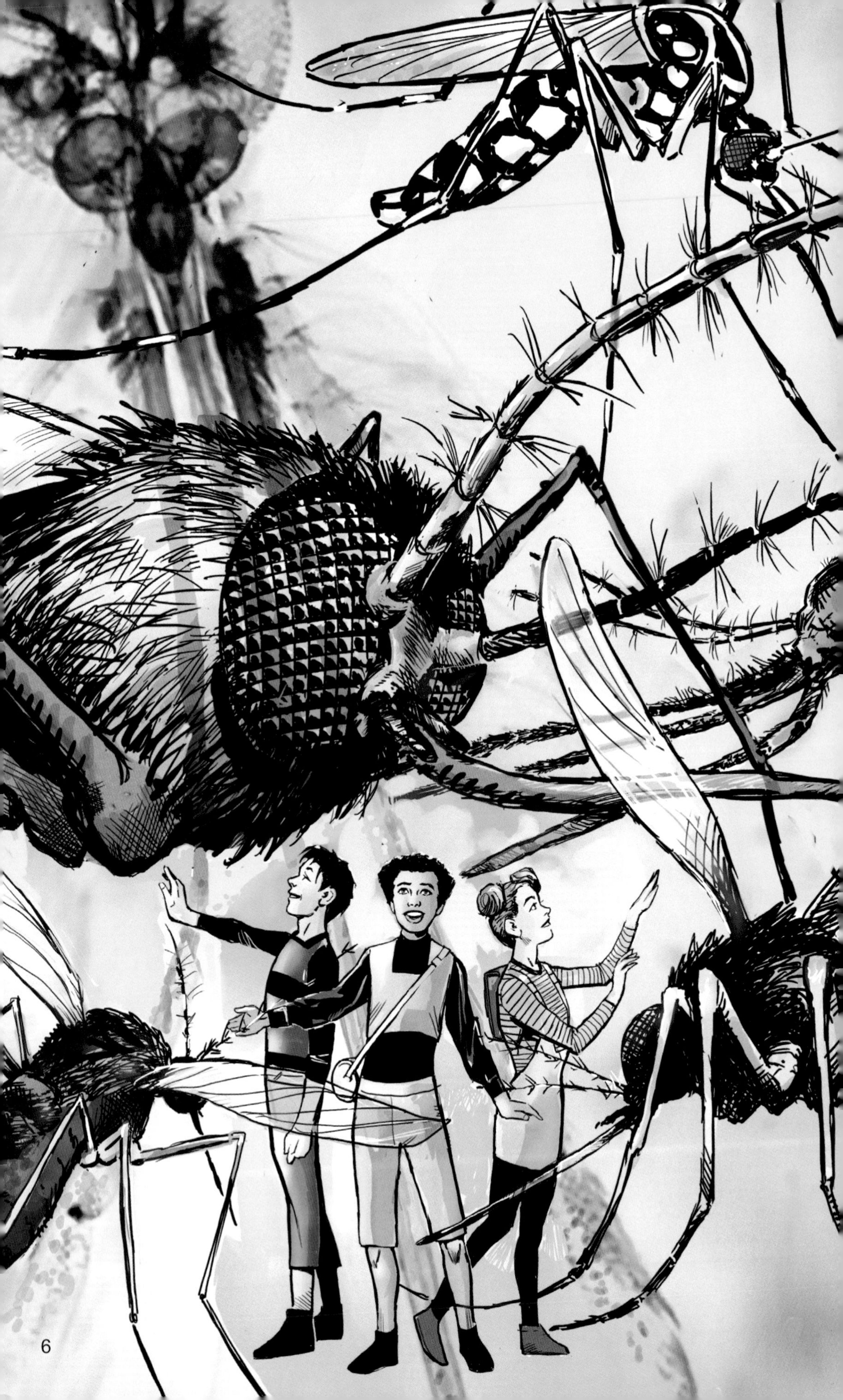

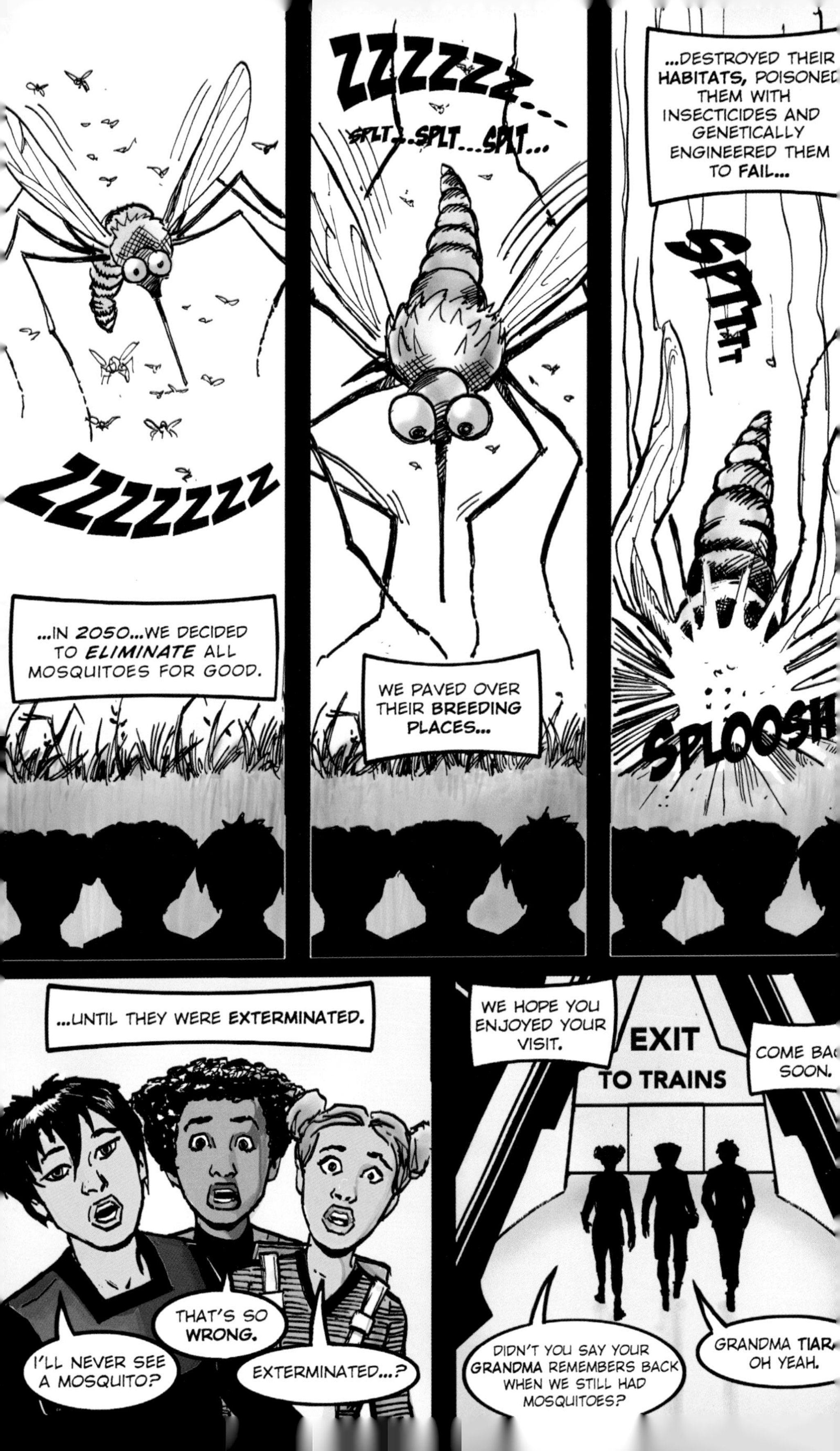
ZZZZZZZ
...IN 2050...WE DECIDED TO ELIMINATE ALL MOSQUITOES FOR GOOD.
ZZZZZZZ...
SPLT...SPLT...SPLT...
WE PAVED OVER THEIR BREEDING PLACES...
...DESTROYED THEIR HABITATS, POISONED THEM WITH INSECTICIDES AND GENETICALLY ENGINEERED THEM TO FAIL...
SPTTTT
SPLOOSH
...UNTIL THEY WERE EXTERMINATED.
I'LL NEVER SEE A MOSQUITO?
THAT'S SO WRONG.
EXTERMINATED...?
WE HOPE YOU ENJOYED YOUR VISIT.
EXIT
TO TRAINS
COME BAC
SOON.
DIDN'T YOU SAY YOUR GRANDMA REMEMBERS BACK WHEN WE STILL HAD MOSQUITOES?
GRANDMA TIAR
OH YEAH.

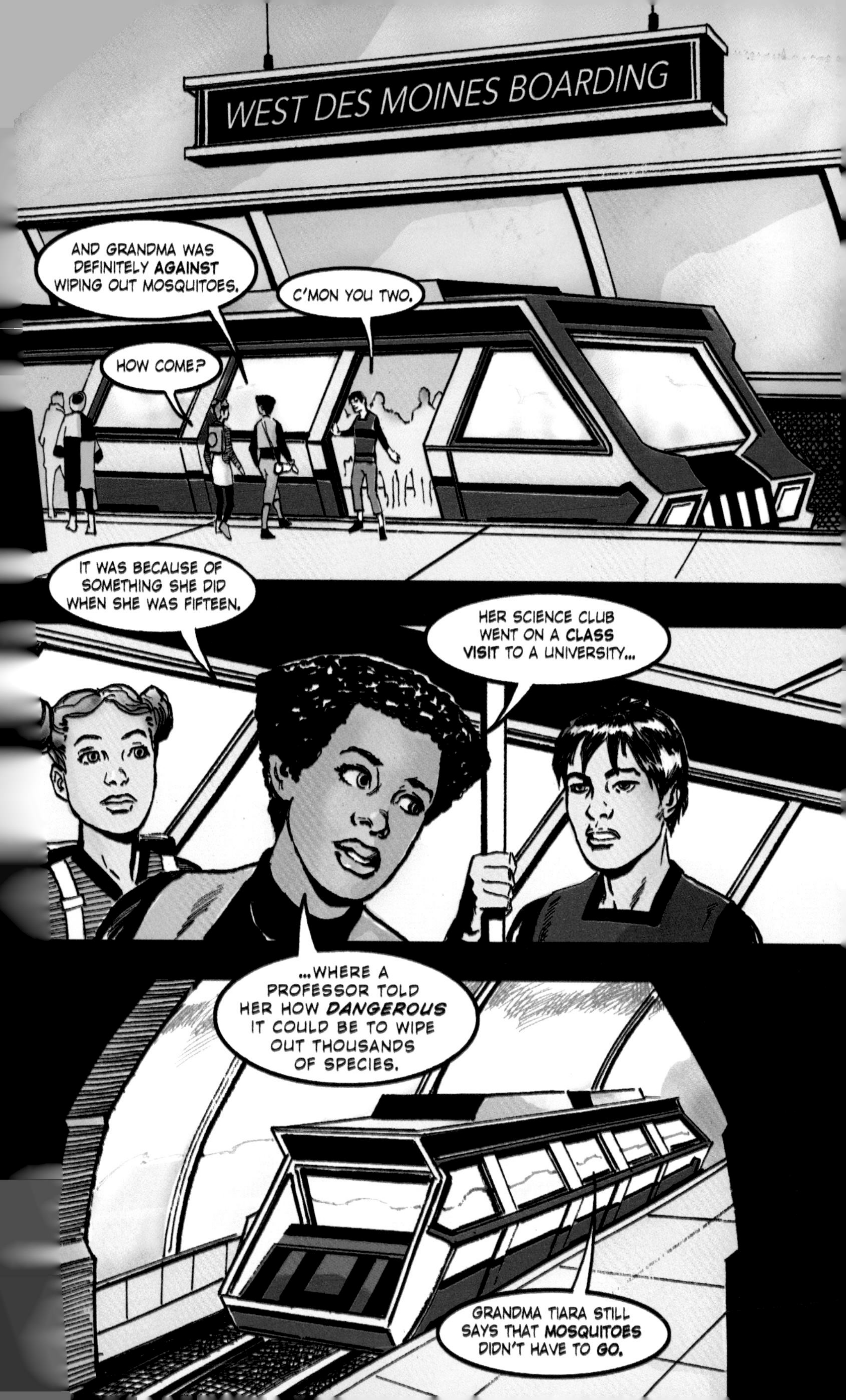
WEST DES MOINES BOARDING
AND GRANDMA WAS DEFINITELY AGAINST WIPING OUT MOSQUITOES.
C'MON YOU TWO.
HOW COME?
IT WAS BECAUSE OF SOMETHING SHE DID WHEN SHE WAS FIFTEEN.
HER SCIENCE CLUB WENT ON A CLASS VISIT TO A UNIVERSITY...
...WHERE A PROFESSOR TOLD HER HOW DANGEROUS IT COULD BE TO WIPE OUT THOUSANDS OF SPECIES.
GRANDMA TIARA STILL SAYS THAT MOSQUITOES DIDN'T HAVE TO GO.

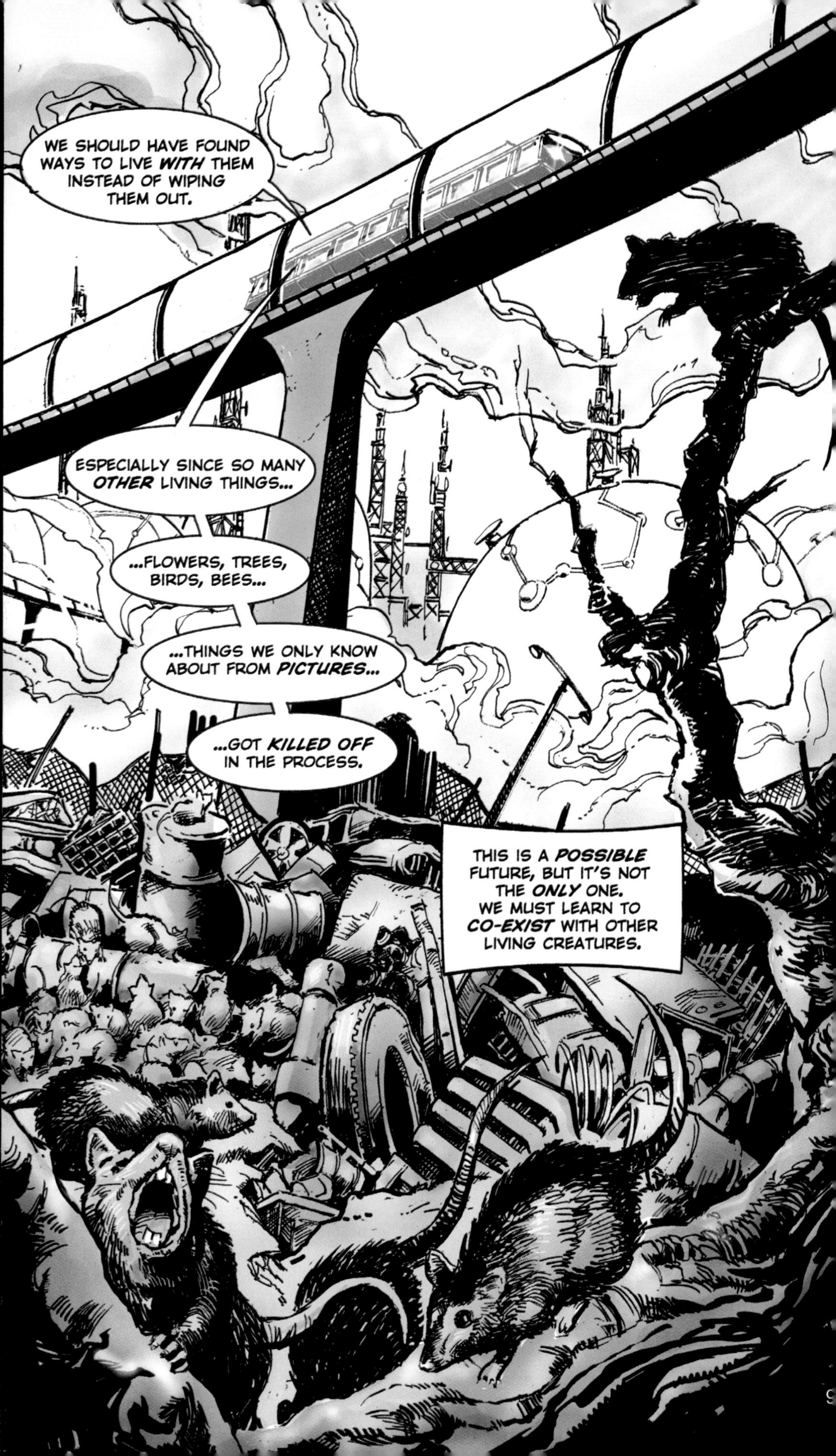
WE SHOULD HAVE FOUND WAYS TO LIVE WITH THEM INSTEAD OF WIPING THEM OUT.
ESPECIALLY SINCE SO MANY OTHER LIVING THINGS...
...FLOWERS, TREES, BIRDS, BEES...
...THINGS WE ONLY KNOW ABOUT FROM PICTURES...
...GOT KILLED OFF IN THE PROCESS.
THIS IS A POSSIBLE FUTURE, BUT IT'S NOT THE ONLY ONE. WE MUST LEARN TO CO-EXIST WITH OTHER LIVING CREATURES.

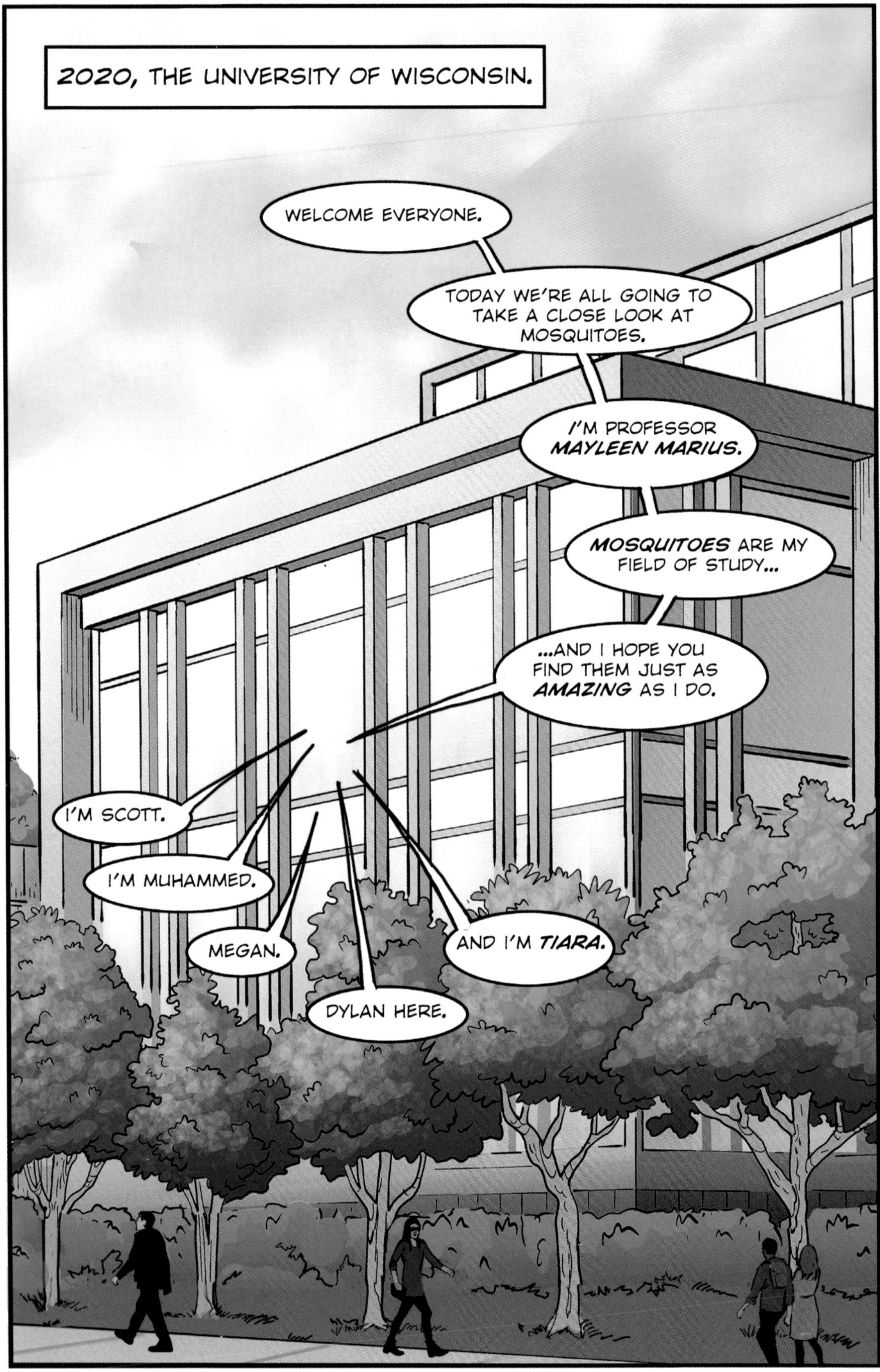
2020, THE UNIVERSITY OF WISCONSIN.
WELCOME EVERYONE.
TODAY WE'RE ALL GOING TO TAKE A CLOSE LOOK AT MOSQUITOES.
I'M PROFESSOR *MAYLEEN MARIUS.*
MOSQUITOES ARE MY FIELD OF STUDY...
...AND I HOPE YOU FIND THEM JUST AS *AMAZING* AS I DO.
I'M SCOTT.
I'M MUHAMMED.
MEGAN.
DYLAN HERE.
AND I'M *TIARA.*

The Hard Life of Mosquito Moms

Story by Bob Hall and Judy Diamond

Illustrated by Michael Cavallaro

Lettering by Mike Edholm

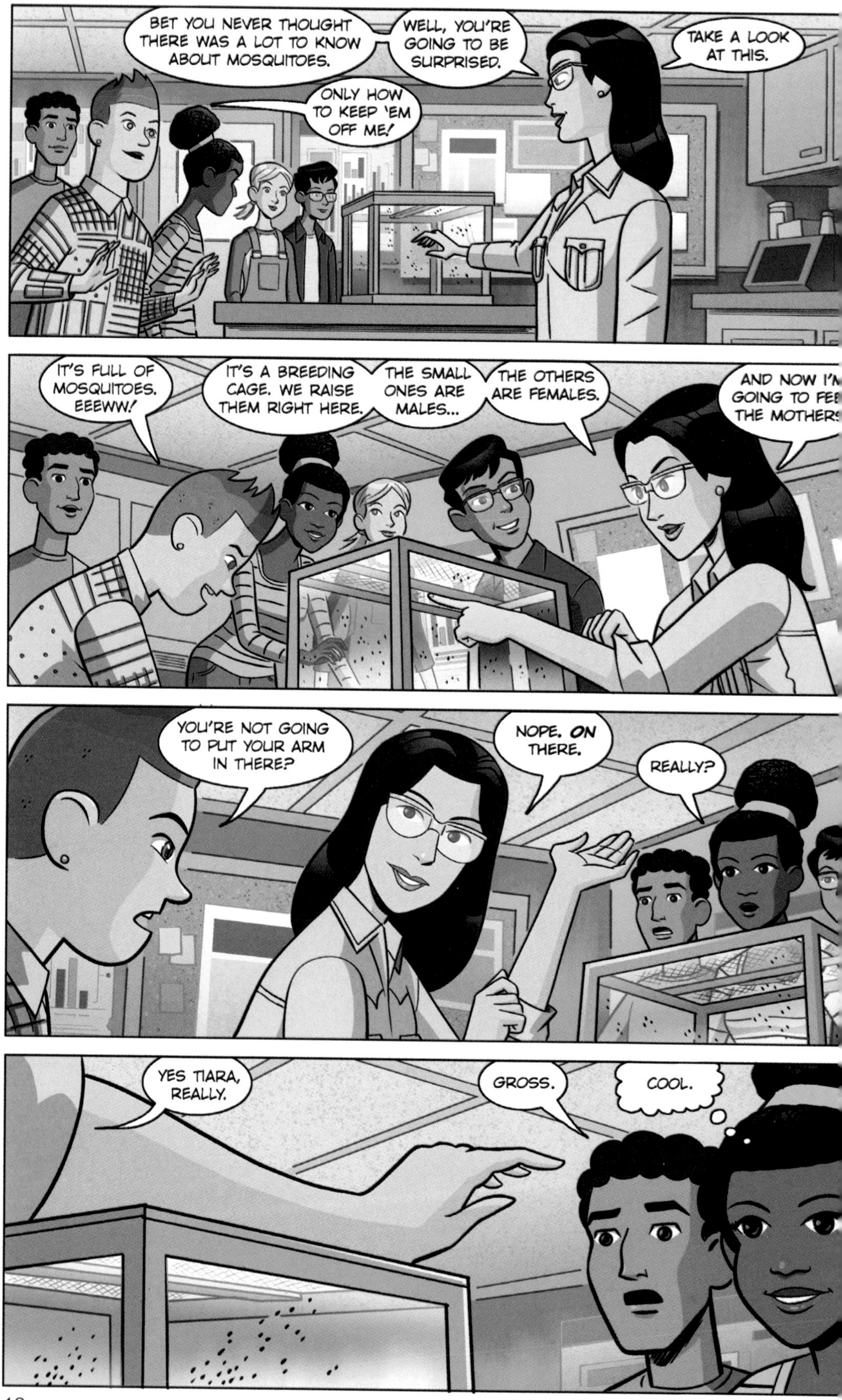

BET YOU NEVER THOUGHT THERE WAS A LOT TO KNOW ABOUT MOSQUITOES.
ONLY HOW TO KEEP 'EM OFF ME!
WELL, YOU'RE GOING TO BE SURPRISED.
TAKE A LOOK AT THIS.
IT'S FULL OF MOSQUITOES. EEEWW!
IT'S A BREEDING CAGE. WE RAISE THEM RIGHT HERE.
THE SMALL ONES ARE MALES...
THE OTHERS ARE FEMALES.
AND NOW I' GOING TO FE THE MOTHER
YOU'RE NOT GOING TO PUT YOUR ARM IN THERE?
NOPE. ***ON*** THERE.
REALLY?
YES TIARA, REALLY.
GROSS.
COOL.

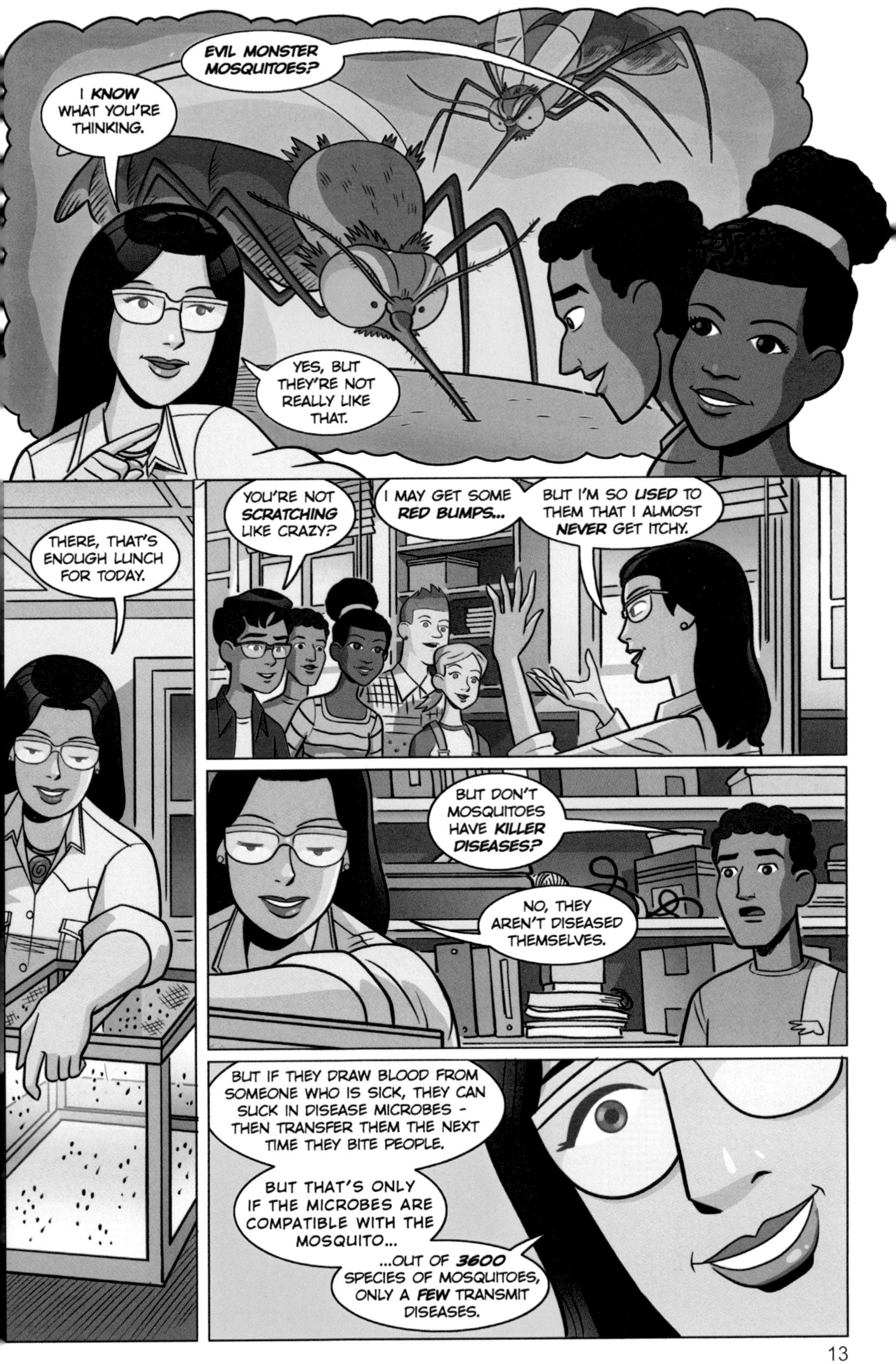
I KNOW WHAT YOU'RE THINKING.
EVIL MONSTER MOSQUITOES?
YES, BUT THEY'RE NOT REALLY LIKE THAT.
THERE, THAT'S ENOUGH LUNCH FOR TODAY.
YOU'RE NOT SCRATCHING LIKE CRAZY?
I MAY GET SOME RED BUMPS...
BUT I'M SO USED TO THEM THAT I ALMOST NEVER GET ITCHY.
BUT DON'T MOSQUITOES HAVE KILLER DISEASES?
NO, THEY AREN'T DISEASED THEMSELVES.
BUT IF THEY DRAW BLOOD FROM SOMEONE WHO IS SICK, THEY CAN SUCK IN DISEASE MICROBES - THEN TRANSFER THEM THE NEXT TIME THEY BITE PEOPLE.
BUT THAT'S ONLY IF THE MICROBES ARE COMPATIBLE WITH THE MOSQUITO...
...OUT OF 3600 SPECIES OF MOSQUITOES, ONLY A FEW TRANSMIT DISEASES.

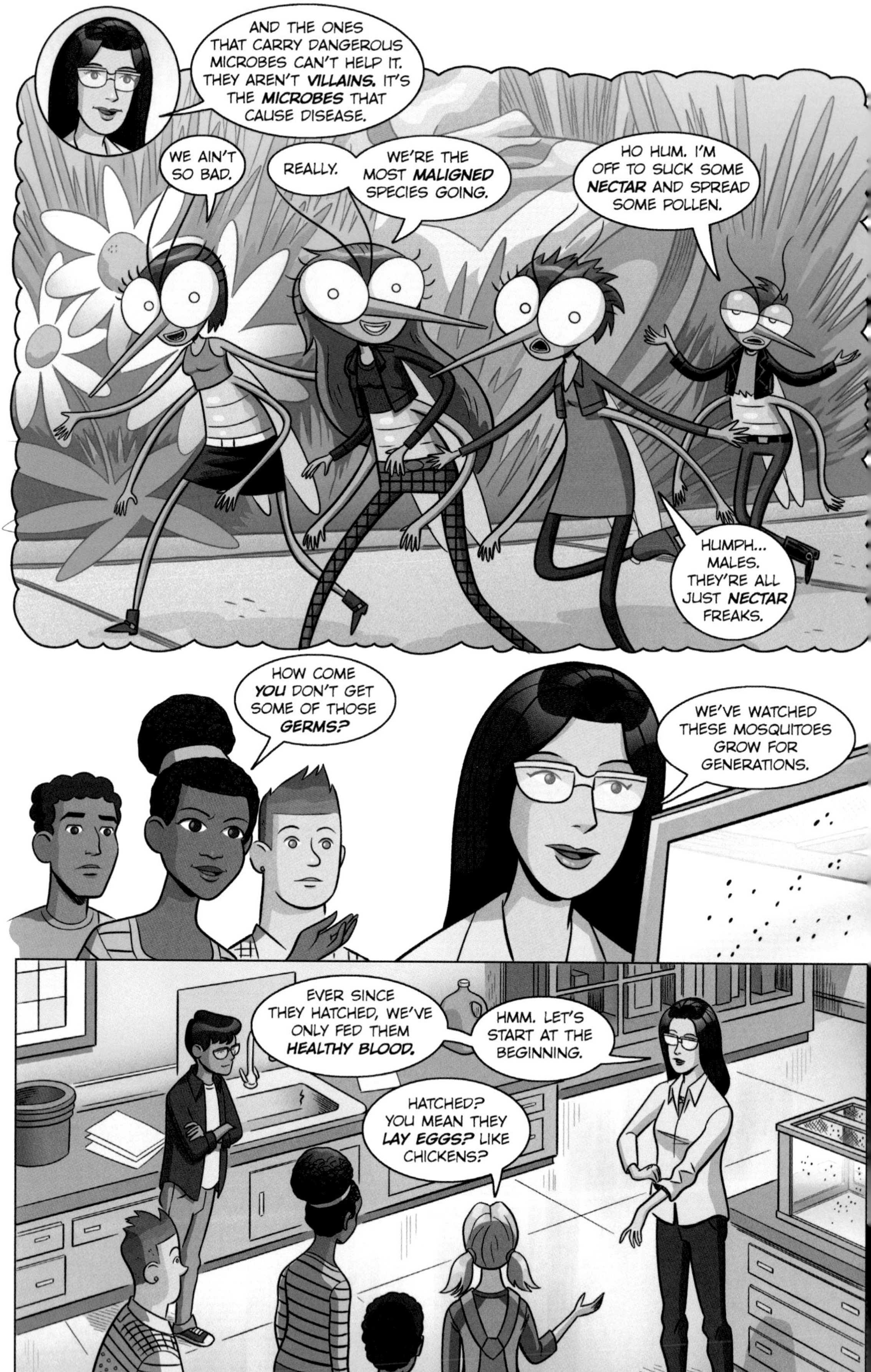
AND THE ONES THAT CARRY DANGEROUS MICROBES CAN'T HELP IT. THEY AREN'T VILLAINS. IT'S THE MICROBES THAT CAUSE DISEASE.
WE AIN'T SO BAD.
REALLY.
WE'RE THE MOST MALIGNED SPECIES GOING.
HO HUM. I'M OFF TO SUCK SOME NECTAR AND SPREAD SOME POLLEN.
HUMPH... MALES. THEY'RE ALL JUST NECTAR FREAKS.
HOW COME YOU DON'T GET SOME OF THOSE GERMS?
WE'VE WATCHED THESE MOSQUITOES GROW FOR GENERATIONS.
EVER SINCE THEY HATCHED, WE'VE ONLY FED THEM HEALTHY BLOOD.
HMM. LET'S START AT THE BEGINNING.
HATCHED? YOU MEAN THEY LAY EGGS? LIKE CHICKENS?

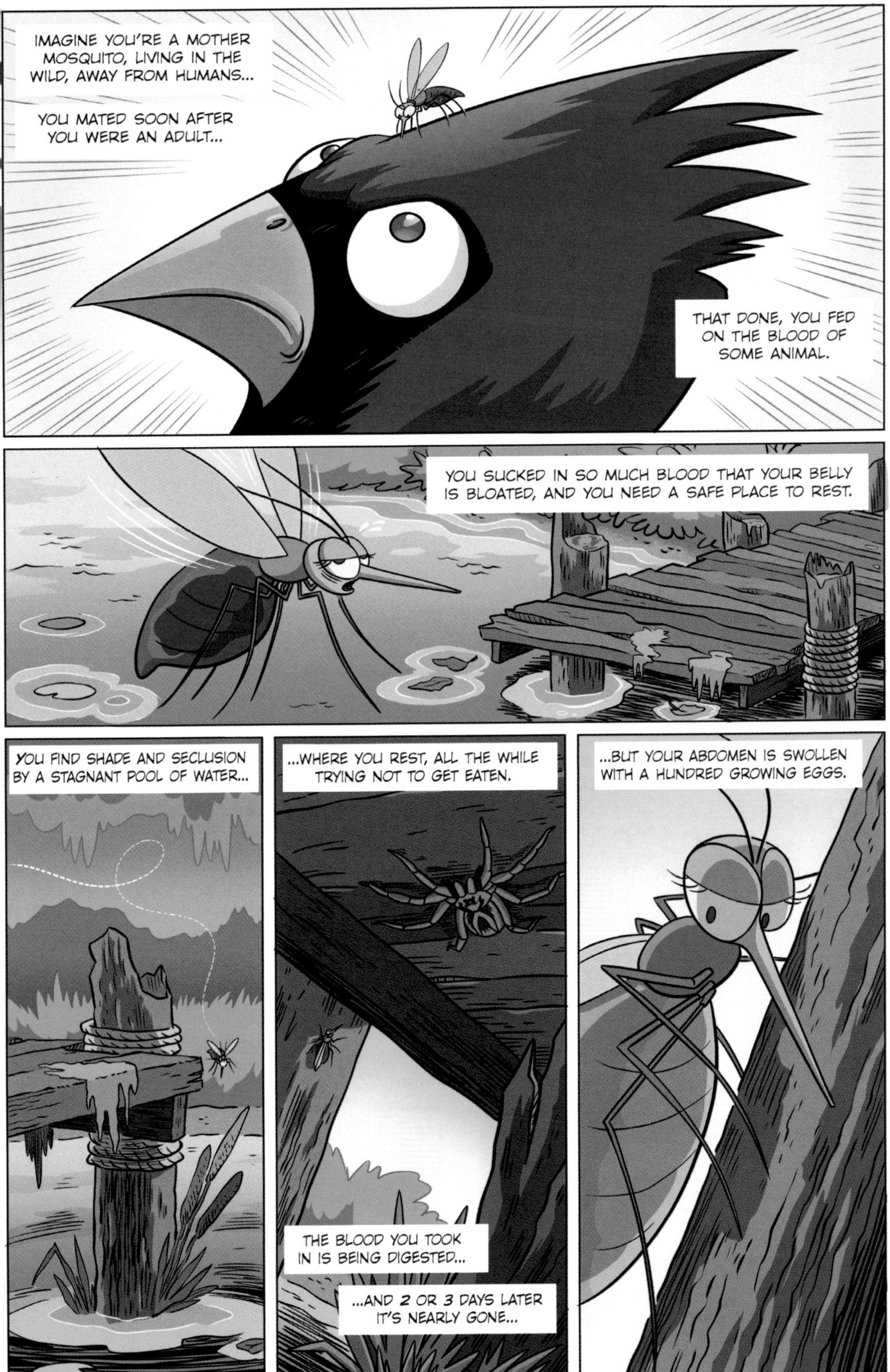
IMAGINE YOU'RE A MOTHER MOSQUITO, LIVING IN THE WILD, AWAY FROM HUMANS...
YOU MATED SOON AFTER YOU WERE AN ADULT...
THAT DONE, YOU FED ON THE BLOOD OF SOME ANIMAL.
YOU SUCKED IN SO MUCH BLOOD THAT YOUR BELLY IS BLOATED, AND YOU NEED A SAFE PLACE TO REST.
YOU FIND SHADE AND SECLUSION BY A STAGNANT POOL OF WATER...
...WHERE YOU REST, ALL THE WHILE TRYING NOT TO GET EATEN.
THE BLOOD YOU TOOK IN IS BEING DIGESTED...
...AND 2 OR 3 DAYS LATER IT'S NEARLY GONE...
...BUT YOUR ABDOMEN IS SWOLLEN WITH A HUNDRED GROWING EGGS.

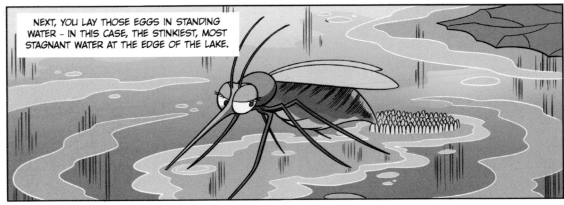

NEXT, YOU LAY THOSE EGGS IN STANDING WATER - IN THIS CASE, THE STINKIEST, MOST STAGNANT WATER AT THE EDGE OF THE LAKE.

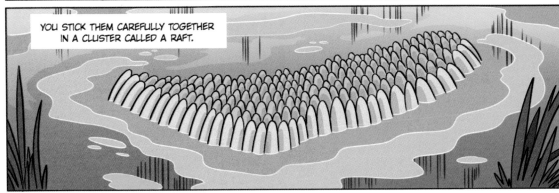

YOU STICK THEM CAREFULLY TOGETHER IN A CLUSTER CALLED A RAFT.

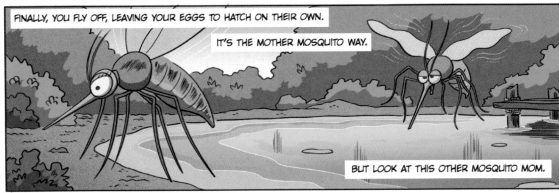

FINALLY, YOU FLY OFF, LEAVING YOUR EGGS TO HATCH ON THEIR OWN.

IT'S THE MOTHER MOSQUITO WAY.

BUT LOOK AT THIS OTHER MOSQUITO MOM.

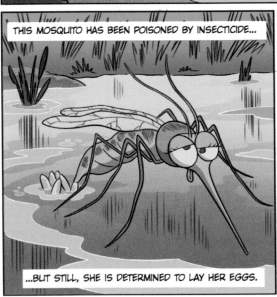

THIS MOSQUITO HAS BEEN POISONED BY INSECTICIDE...

...BUT STILL, SHE IS DETERMINED TO LAY HER EGGS.

LOOK! AT LEAST 100 EGGS IN THE HEALTHY RAFT AND ONLY A FEW IN THE OTHER.

LARVAE THAT HATCH FROM THOSE EGGS WILL BE TOO WEAK TO MAKE IT.

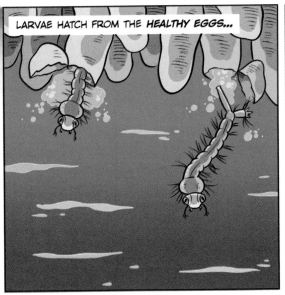

LARVAE HATCH FROM THE *HEALTHY EGGS*...

...AND CLING JUST BELOW THE SURFACE OF THE POND...

...BREATHING AIR THROUGH SIPHON TUBES...

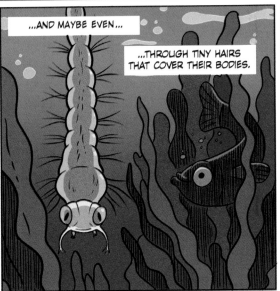

...AND MAYBE EVEN...

...THROUGH TINY HAIRS THAT COVER THEIR BODIES.

AT THIS STAGE, THEY ARE EASY PREY TO FISH AND OTHER PREDATORS.

SOME ESCAPE...

SOME DON'T...

HELP... ME...

EEEEEEK...

OK, I KNOW, LARVAE CAN'T SCREAM - BUT YOU GET THE IDEA.

ONLY A FEW LARVAE SURVIVE.

EVENTUALLY THOSE THAT DO TRANSFORM INTO THE *PUPAL STAGE*.

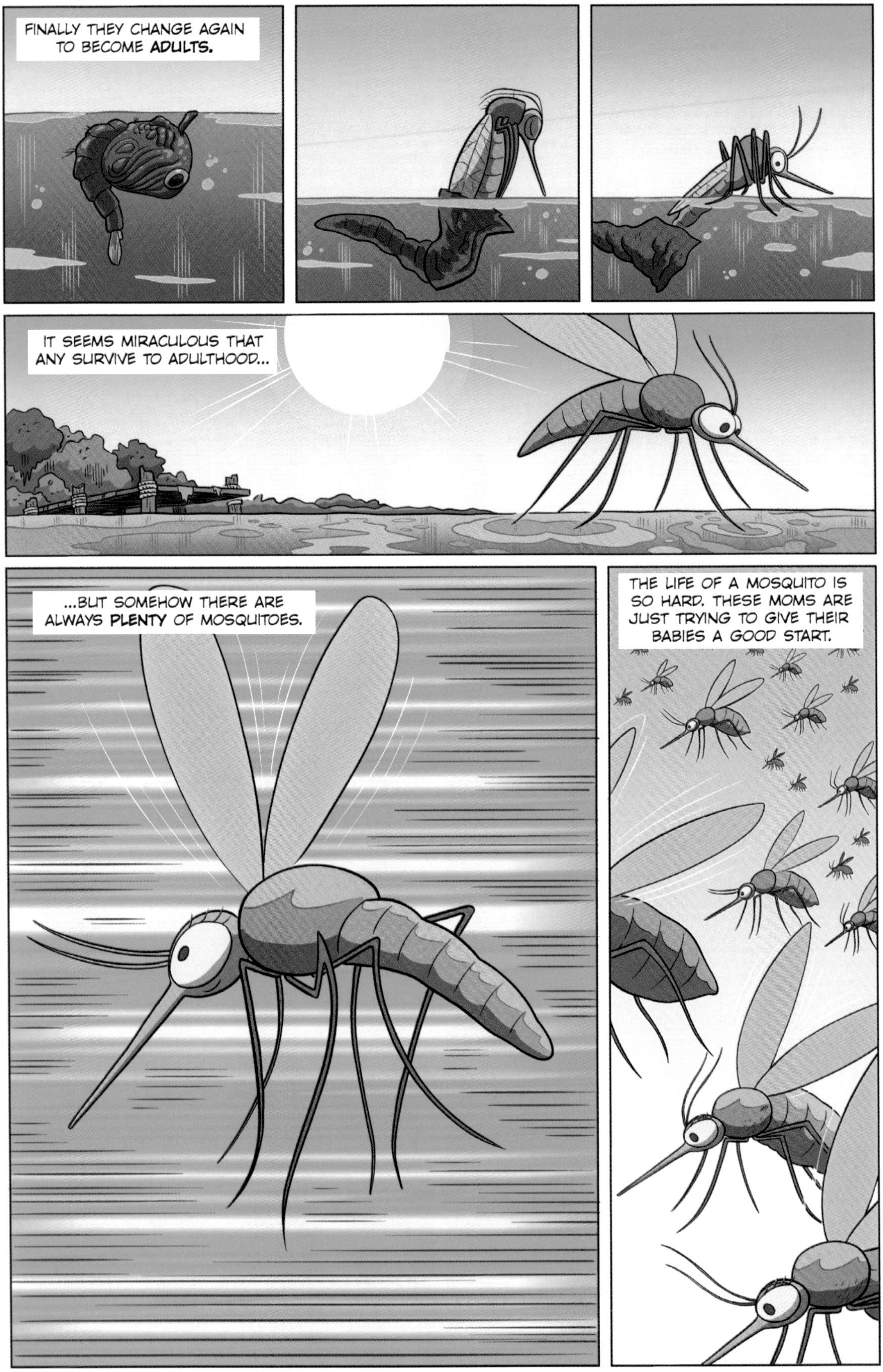

FINALLY THEY CHANGE AGAIN TO BECOME **ADULTS.**
IT SEEMS MIRACULOUS THAT ANY SURVIVE TO ADULTHOOD...
...BUT SOMEHOW THERE ARE ALWAYS **PLENTY** OF MOSQUITOES.
THE LIFE OF A MOSQUITO IS SO HARD. THESE MOMS ARE JUST TRYING TO GIVE THEIR BABIES A GOOD START.

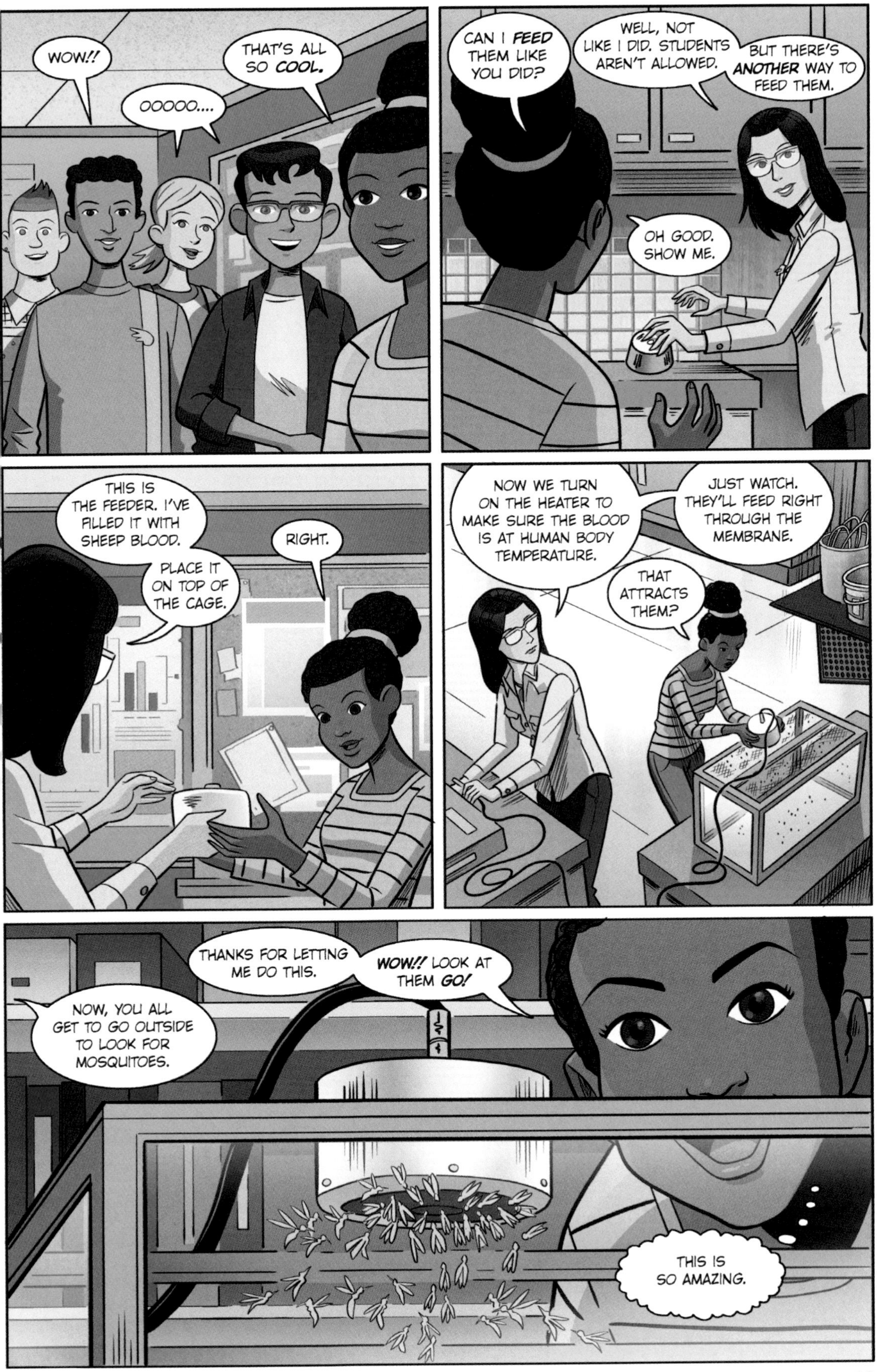
WOW!!
OOOOO....
THAT'S ALL SO **COOL.**
CAN I **FEED** THEM LIKE YOU DID?
WELL, NOT LIKE I DID. STUDENTS AREN'T ALLOWED.
BUT THERE'S **ANOTHER** WAY TO FEED THEM.
OH GOOD. SHOW ME.
THIS IS THE FEEDER. I'VE FILLED IT WITH SHEEP BLOOD.
PLACE IT ON TOP OF THE CAGE.
RIGHT.
NOW WE TURN ON THE HEATER TO MAKE SURE THE BLOOD IS AT HUMAN BODY TEMPERATURE.
JUST WATCH. THEY'LL FEED RIGHT THROUGH THE MEMBRANE.
THAT ATTRACTS THEM?
THANKS FOR LETTING ME DO THIS.
WOW!! LOOK AT THEM **GO!**
NOW, YOU ALL GET TO GO OUTSIDE TO LOOK FOR MOSQUITOES.
THIS IS SO AMAZING.

Ridding the World of Pesky Mosquitoes

Story by Bob Hall and Judy Diamond

Illustrated by Bob Camp

Lettering by Mike Edholm

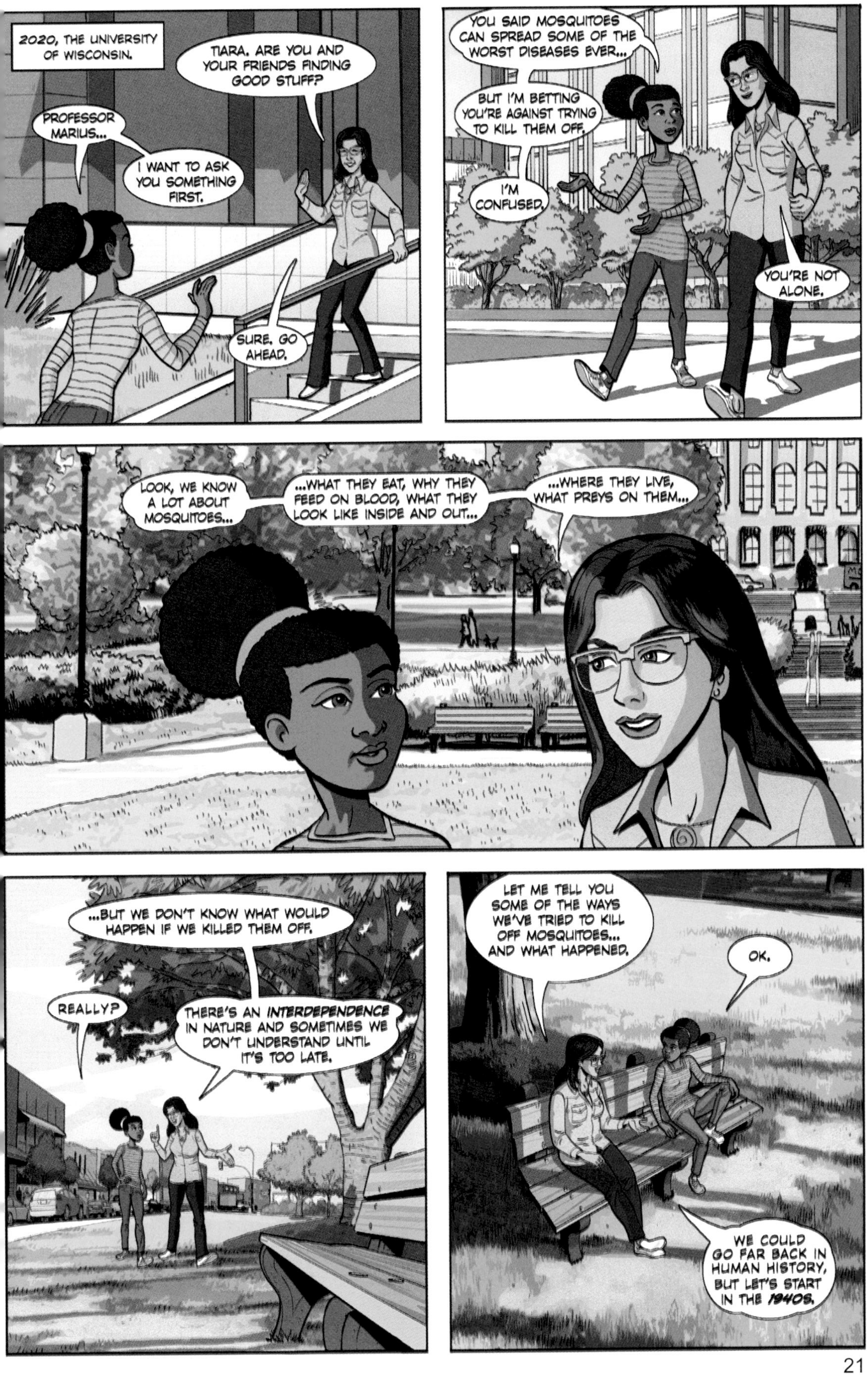
2020, THE UNIVERSITY OF WISCONSIN.
TIARA. ARE YOU AND YOUR FRIENDS FINDING GOOD STUFF?
PROFESSOR MARIUS...
I WANT TO ASK YOU SOMETHING FIRST.
SURE. GO AHEAD.
YOU SAID MOSQUITOES CAN SPREAD SOME OF THE WORST DISEASES EVER...
BUT I'M BETTING YOU'RE AGAINST TRYING TO KILL THEM OFF.
I'M CONFUSED.
YOU'RE NOT ALONE.
LOOK, WE KNOW A LOT ABOUT MOSQUITOES...
...WHAT THEY EAT, WHY THEY FEED ON BLOOD, WHAT THEY LOOK LIKE INSIDE AND OUT...
...WHERE THEY LIVE, WHAT PREYS ON THEM...
...BUT WE DON'T KNOW WHAT WOULD HAPPEN IF WE KILLED THEM OFF.
REALLY?
THERE'S AN *INTERDEPENDENCE* IN NATURE AND SOMETIMES WE DON'T UNDERSTAND UNTIL IT'S TOO LATE.
LET ME TELL YOU SOME OF THE WAYS WE'VE TRIED TO KILL OFF MOSQUITOES... AND WHAT HAPPENED.
OK.
WE COULD GO FAR BACK IN HUMAN HISTORY, BUT LET'S START IN THE *1940S*.

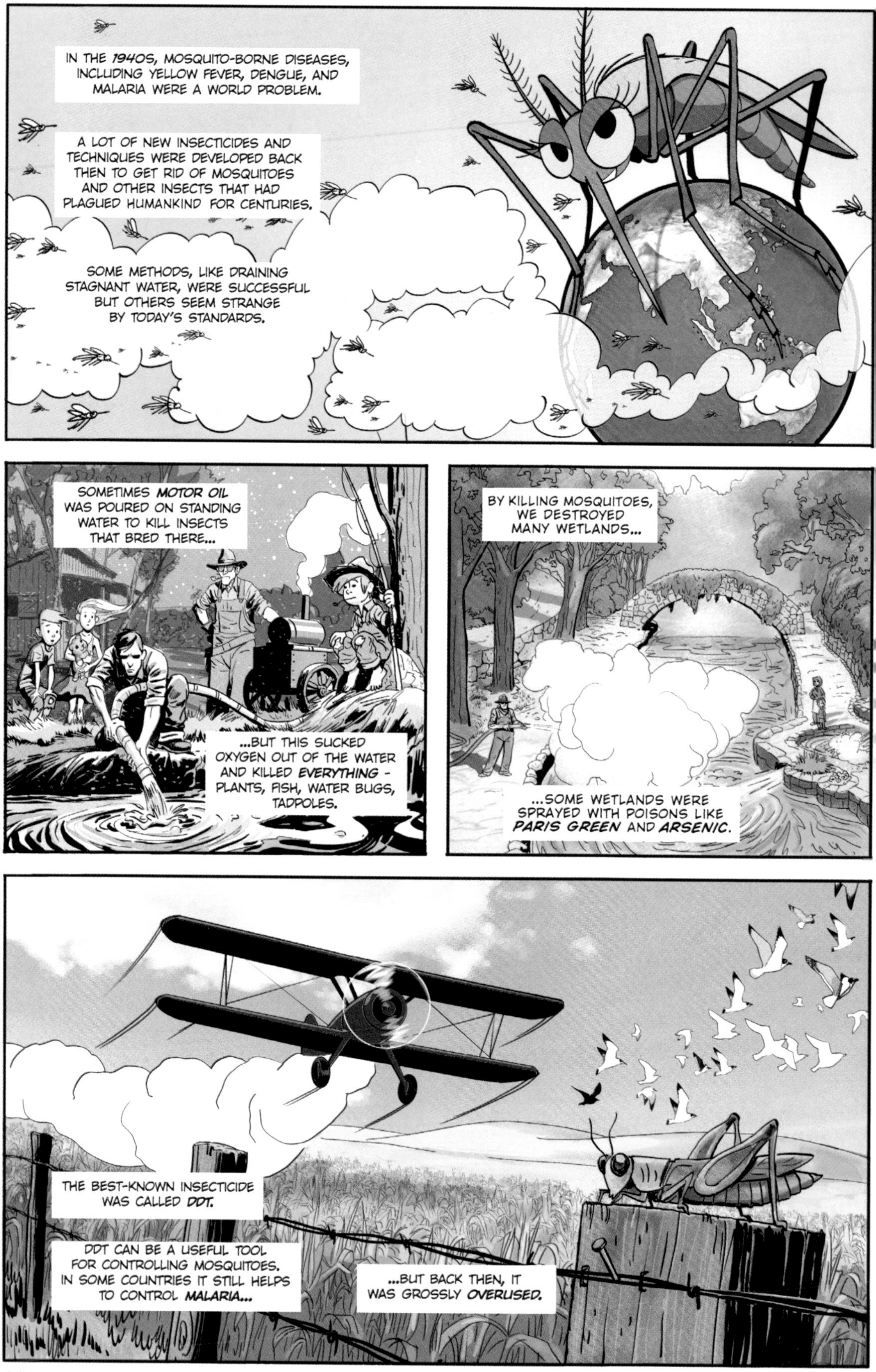
IN THE 1940S, MOSQUITO-BORNE DISEASES, INCLUDING YELLOW FEVER, DENGUE, AND MALARIA WERE A WORLD PROBLEM.
A LOT OF NEW INSECTICIDES AND TECHNIQUES WERE DEVELOPED BACK THEN TO GET RID OF MOSQUITOES AND OTHER INSECTS THAT HAD PLAGUED HUMANKIND FOR CENTURIES.
SOME METHODS, LIKE DRAINING STAGNANT WATER, WERE SUCCESSFUL BUT OTHERS SEEM STRANGE BY TODAY'S STANDARDS.
SOMETIMES MOTOR OIL WAS POURED ON STANDING WATER TO KILL INSECTS THAT BRED THERE...
...BUT THIS SUCKED OXYGEN OUT OF THE WATER AND KILLED EVERYTHING - PLANTS, FISH, WATER BUGS, TADPOLES.
BY KILLING MOSQUITOES, WE DESTROYED MANY WETLANDS...
...SOME WETLANDS WERE SPRAYED WITH POISONS LIKE PARIS GREEN AND ARSENIC.
THE BEST-KNOWN INSECTICIDE WAS CALLED DDT.
DDT CAN BE A USEFUL TOOL FOR CONTROLLING MOSQUITOES. IN SOME COUNTRIES IT STILL HELPS TO CONTROL MALARIA...
...BUT BACK THEN, IT WAS GROSSLY OVERUSED.

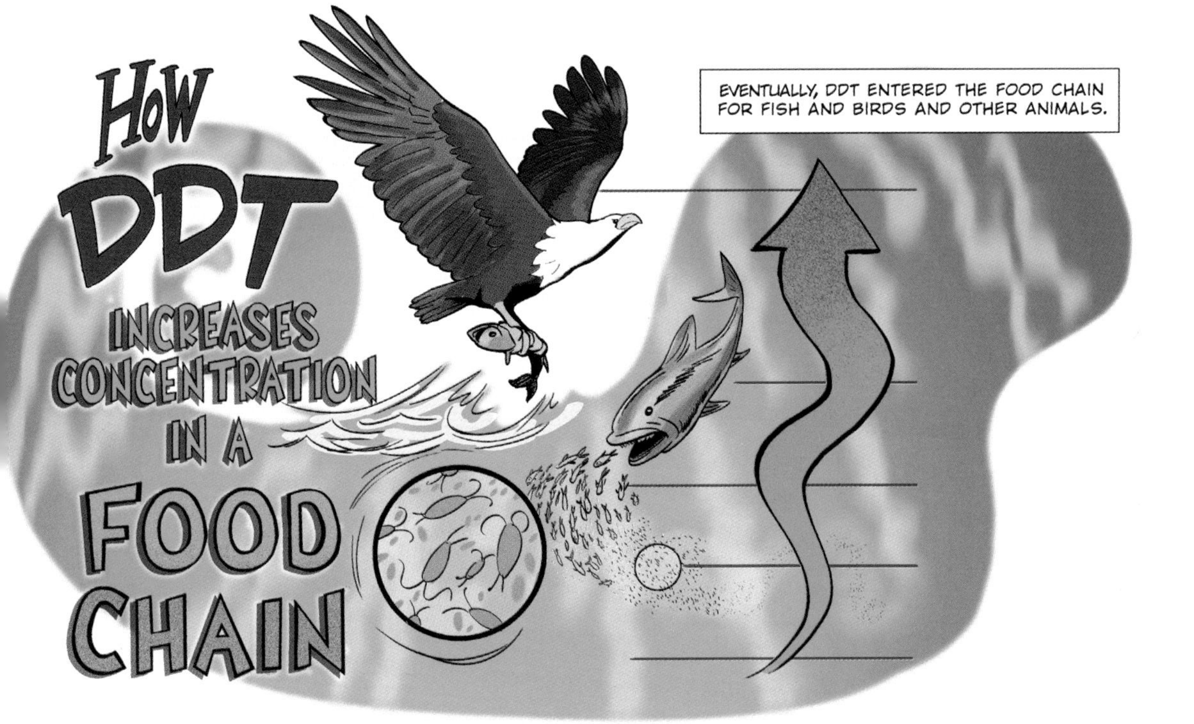
How
DDT
INCREASES
CONCENTRATION
IN A
FOOD
CHAIN
EVENTUALLY, DDT ENTERED THE FOOD CHAIN FOR FISH AND BIRDS AND OTHER ANIMALS.

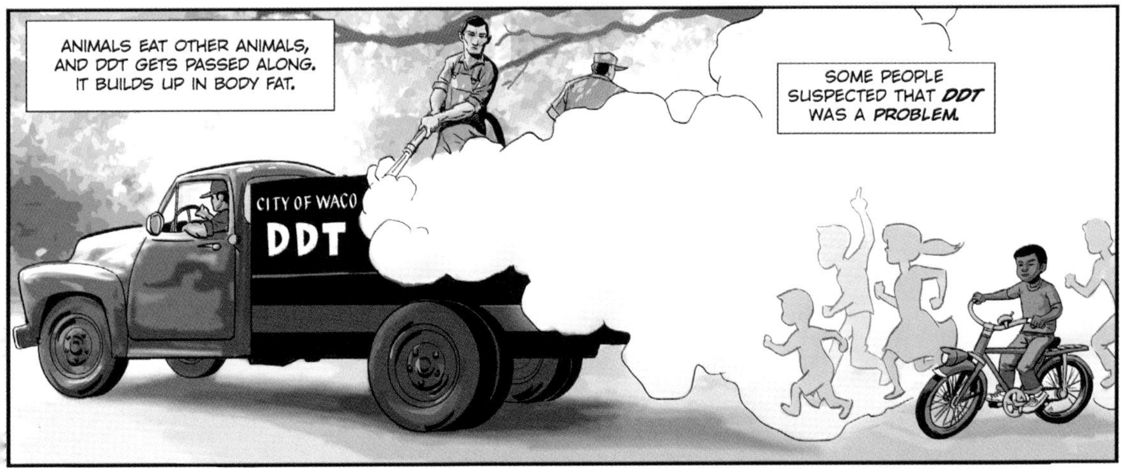
ANIMALS EAT OTHER ANIMALS, AND DDT GETS PASSED ALONG. IT BUILDS UP IN BODY FAT.
CITY OF WACO
DDT
SOME PEOPLE SUSPECTED THAT DDT WAS A PROBLEM.

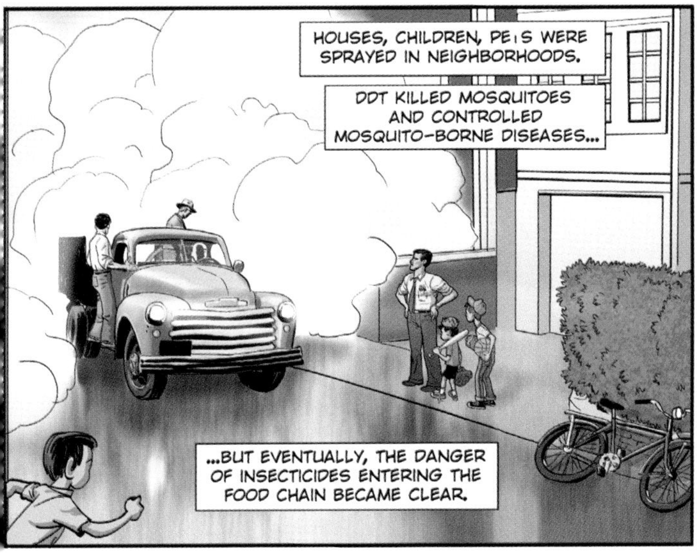
HOUSES, CHILDREN, PETS WERE SPRAYED IN NEIGHBORHOODS.
DDT KILLED MOSQUITOES AND CONTROLLED MOSQUITO-BORNE DISEASES...
...BUT EVENTUALLY, THE DANGER OF INSECTICIDES ENTERING THE FOOD CHAIN BECAME CLEAR.

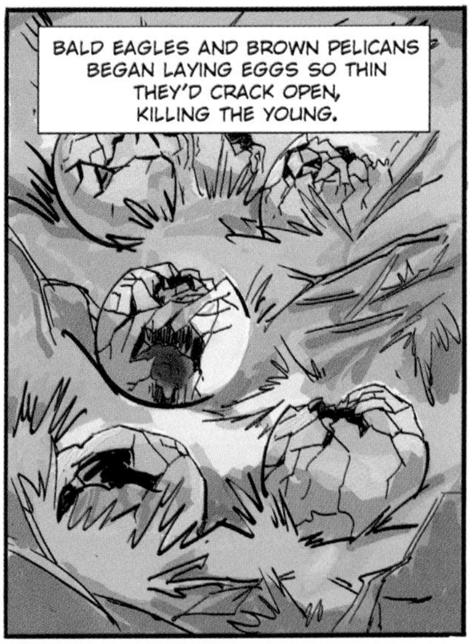
BALD EAGLES AND BROWN PELICANS BEGAN LAYING EGGS SO THIN THEY'D CRACK OPEN, KILLING THE YOUNG.

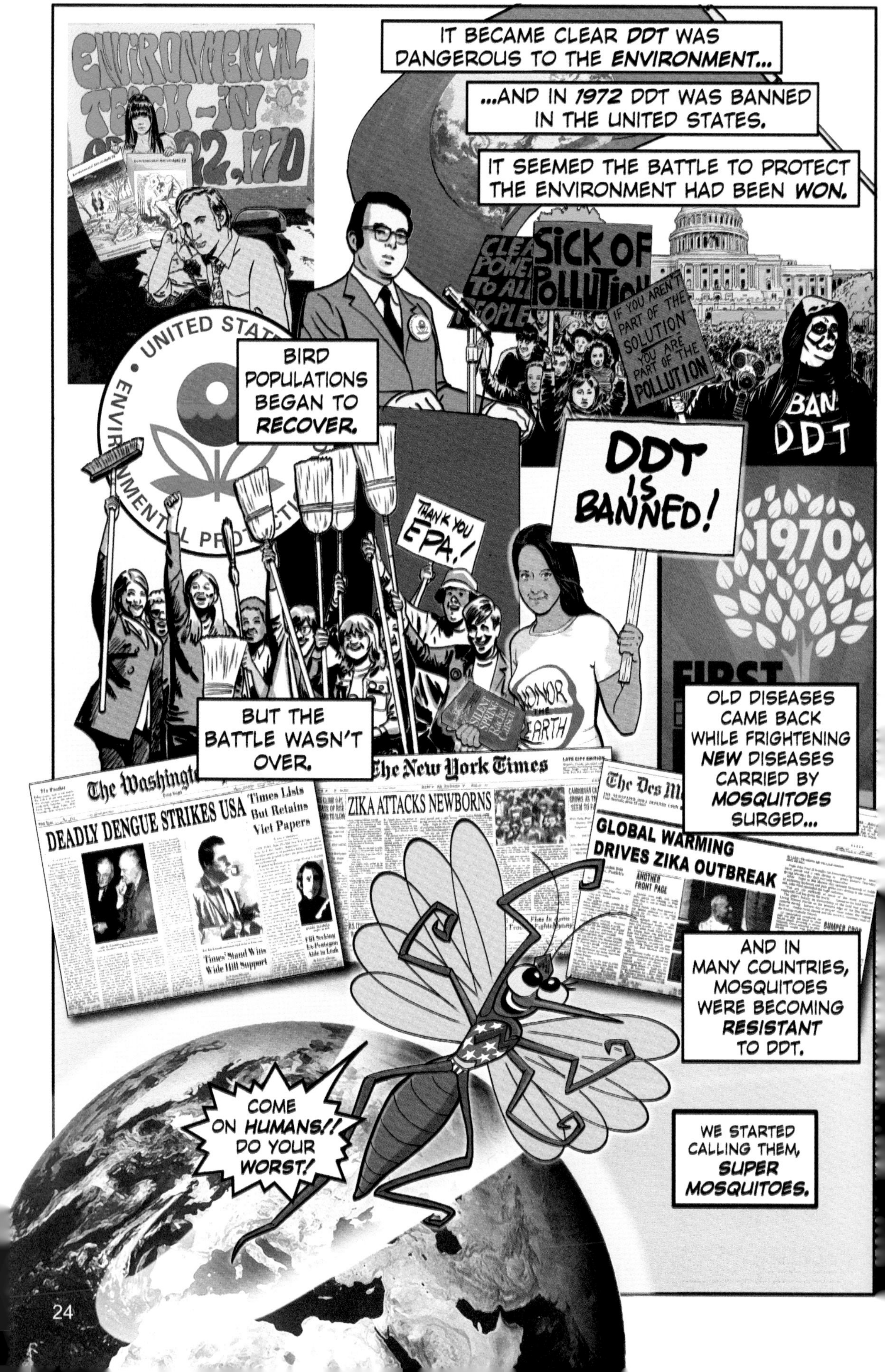

IT BECAME CLEAR DDT WAS DANGEROUS TO THE ENVIRONMENT...
...AND IN 1972 DDT WAS BANNED IN THE UNITED STATES.
IT SEEMED THE BATTLE TO PROTECT THE ENVIRONMENT HAD BEEN WON.
ENVIRONMENTAL TEACH-IN
22, 1970
SICK OF POLLUTION
IF YOU AREN'T PART OF THE SOLUTION YOU ARE PART OF THE POLLUTION
BAN DDT
UNITED STATES
ENVIRONMENTAL PROTECTION
BIRD POPULATIONS BEGAN TO RECOVER.
DDT IS BANNED!
1970
THANK YOU EPA!
HONOR THE EARTH
SILENT SPRING Rachel Carson
BUT THE BATTLE WASN'T OVER.
OLD DISEASES CAME BACK WHILE FRIGHTENING NEW DISEASES CARRIED BY MOSQUITOES SURGED...
The New York Times
ZIKA ATTACKS NEWBORNS
DEADLY DENGUE STRIKES USA
Times Lists But Retains Viet Papers
GLOBAL WARMING DRIVES ZIKA OUTBREAK
AND IN MANY COUNTRIES, MOSQUITOES WERE BECOMING RESISTANT TO DDT.
COME ON HUMANS!! DO YOUR WORST!
WE STARTED CALLING THEM, SUPER MOSQUITOES.

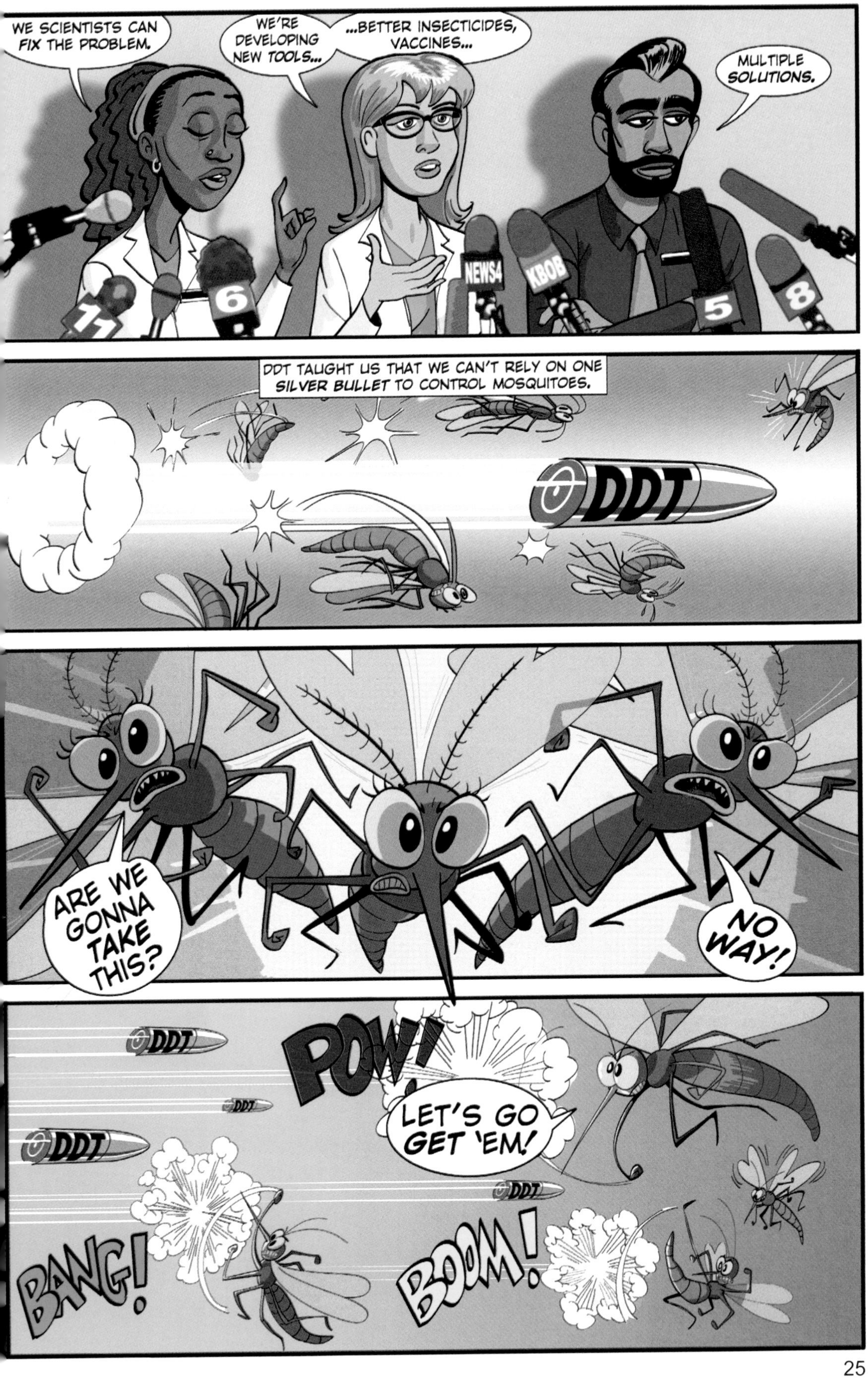
WE SCIENTISTS CAN FIX THE PROBLEM.
WE'RE DEVELOPING NEW TOOLS...
...BETTER INSECTICIDES, VACCINES...
MULTIPLE SOLUTIONS.
NEWS4
KBOB
6
5
8
DDT TAUGHT US THAT WE CAN'T RELY ON ONE SILVER BULLET TO CONTROL MOSQUITOES.
DDT
ARE WE GONNA TAKE THIS?
NO WAY!
POW!
LET'S GO GET 'EM!
DDT
DDT
DDT
DDT
BANG!
BOOM!

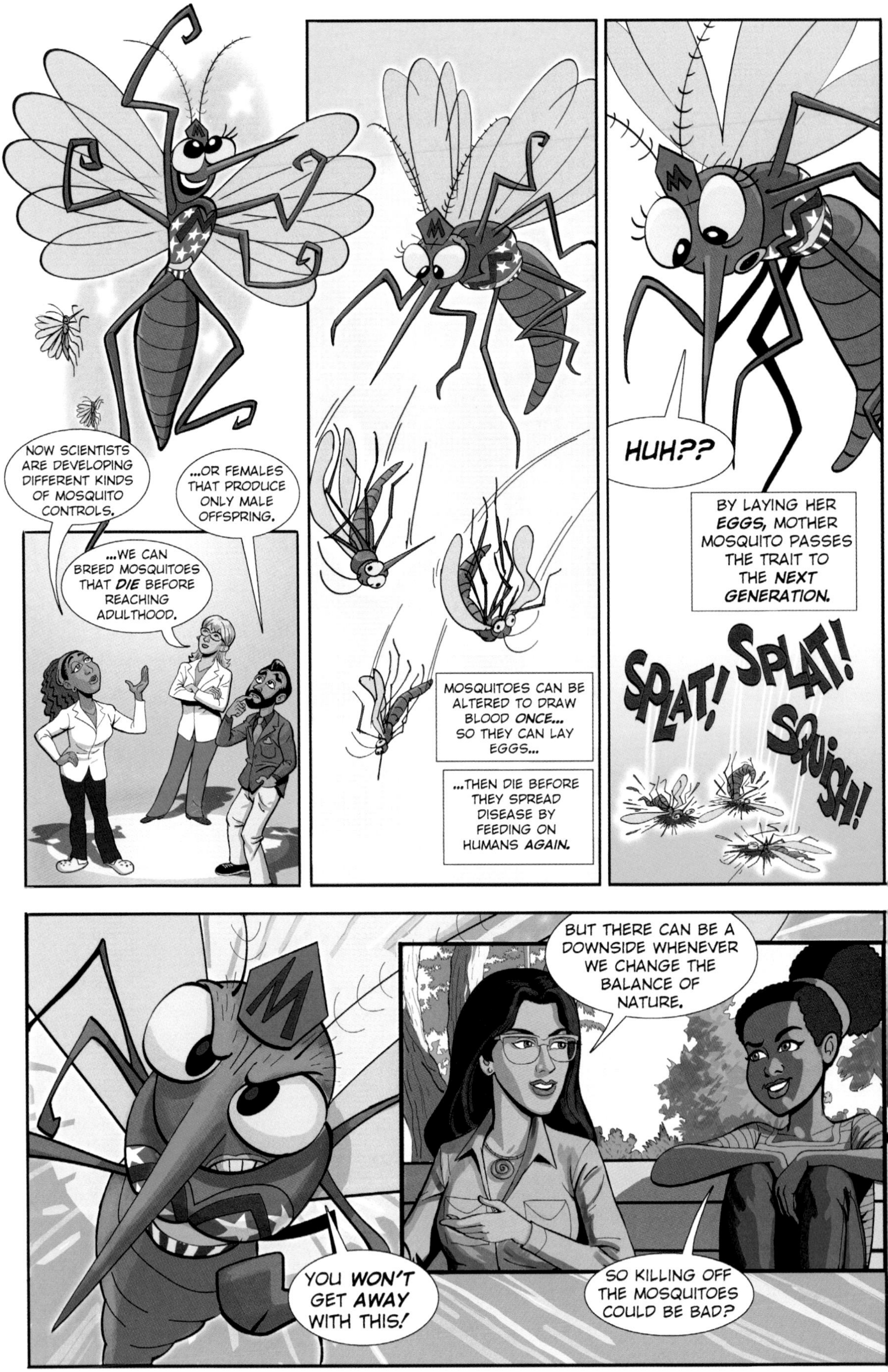
NOW SCIENTISTS ARE DEVELOPING DIFFERENT KINDS OF MOSQUITO CONTROLS.
...OR FEMALES THAT PRODUCE ONLY MALE OFFSPRING.
...WE CAN BREED MOSQUITOES THAT DIE BEFORE REACHING ADULTHOOD.
MOSQUITOES CAN BE ALTERED TO DRAW BLOOD ONCE... SO THEY CAN LAY EGGS...
...THEN DIE BEFORE THEY SPREAD DISEASE BY FEEDING ON HUMANS AGAIN.
HUH??
BY LAYING HER EGGS, MOTHER MOSQUITO PASSES THE TRAIT TO THE NEXT GENERATION.
SPLAT! SPLAT! SQUISH!
YOU WON'T GET AWAY WITH THIS!
BUT THERE CAN BE A DOWNSIDE WHENEVER WE CHANGE THE BALANCE OF NATURE.
SO KILLING OFF THE MOSQUITOES COULD BE BAD?

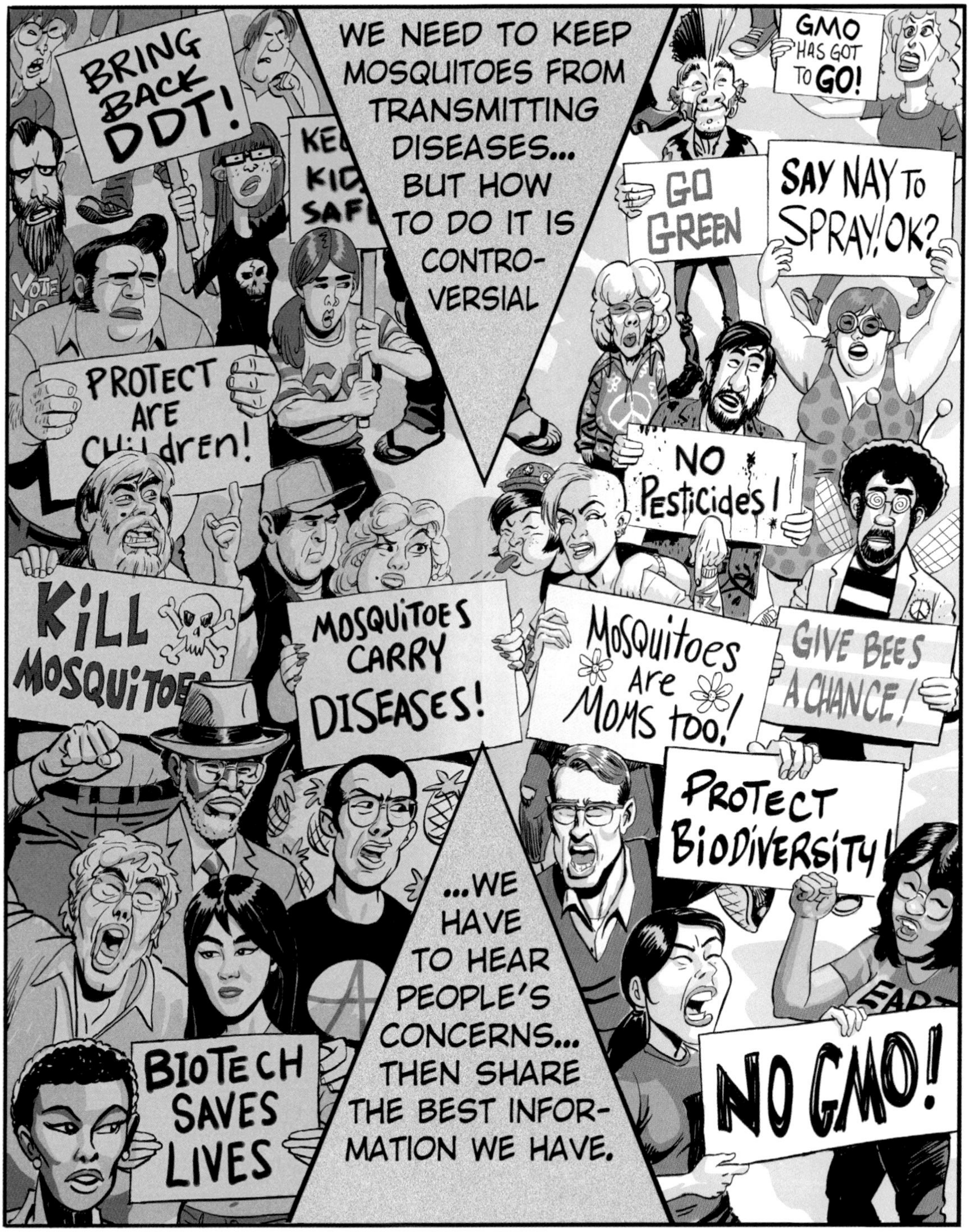
WE NEED TO KEEP MOSQUITOES FROM TRANSMITTING DISEASES... BUT HOW TO DO IT IS CONTRO-VERSIAL
BRING BACK DDT!
GMO HAS GOT TO GO!
GO GREEN
SAY NAY TO SPRAY! OK?
PROTECT ARE CHildren!
NO Pesticides!
KILL MOSQUITOES
MOSQUITOES CARRY DISEASES!
Mosquitoes Are MOMS too!
GIVE BEES A CHANCE!
PROTECT BIODIVERSITY!
BIOTECH SAVES LIVES
NO GMO!
...WE HAVE TO HEAR PEOPLE'S CONCERNS... THEN SHARE THE BEST INFOR-MATION WE HAVE.

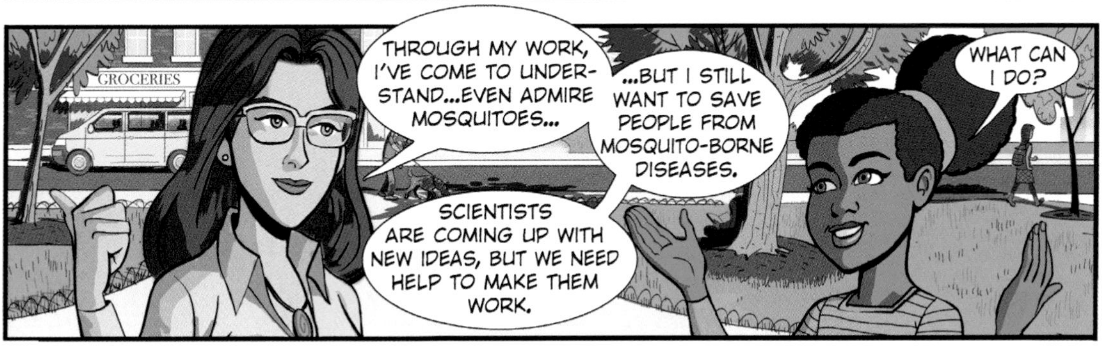
GROCERIES
THROUGH MY WORK, I'VE COME TO UNDER-STAND...EVEN ADMIRE MOSQUITOES...
...BUT I STILL WANT TO SAVE PEOPLE FROM MOSQUITO-BORNE DISEASES.
SCIENTISTS ARE COMING UP WITH NEW IDEAS, BUT WE NEED HELP TO MAKE THEM WORK.
WHAT CAN I DO?

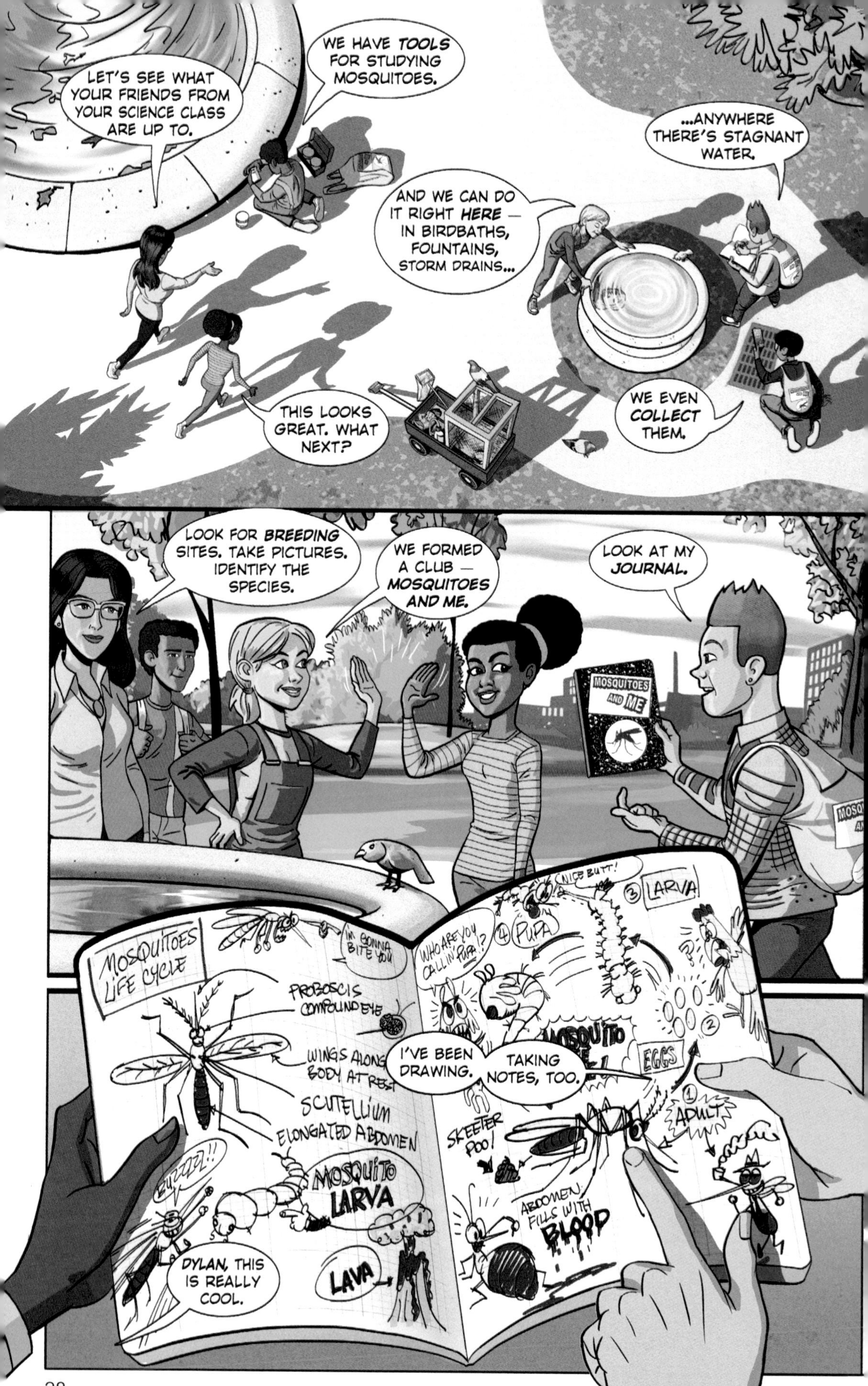
LET'S SEE WHAT YOUR FRIENDS FROM YOUR SCIENCE CLASS ARE UP TO.
WE HAVE *TOOLS* FOR STUDYING MOSQUITOES.
AND WE CAN DO IT RIGHT *HERE* — IN BIRDBATHS, FOUNTAINS, STORM DRAINS...
...ANYWHERE THERE'S STAGNANT WATER.
THIS LOOKS GREAT. WHAT NEXT?
WE EVEN *COLLECT* THEM.
LOOK FOR *BREEDING* SITES. TAKE PICTURES. IDENTIFY THE SPECIES.
WE FORMED A CLUB — *MOSQUITOES AND ME.*
LOOK AT MY *JOURNAL.*
MOSQUITOES AND ME
MOSQUITOES LIFE CYCLE
I'M GONNA BITE YOU
PROBOSCIS
COMPOUND EYE
WINGS ALONG BODY AT REST
SCUTELLIUM
ELONGATED ABDOMEN
BUZZZZ!!
MOSQUITO LARVA
LAVA
NICE BUTT!
3 LARVA
4 PUPA
WHO ARE YOU CALLIN' PUPA!?
2 EGGS
1 ADULT
SKEETER POO!
ABDOMEN FILLS WITH BLOOD
I'VE BEEN DRAWING.
TAKING NOTES, TOO.
DYLAN, THIS IS REALLY COOL.

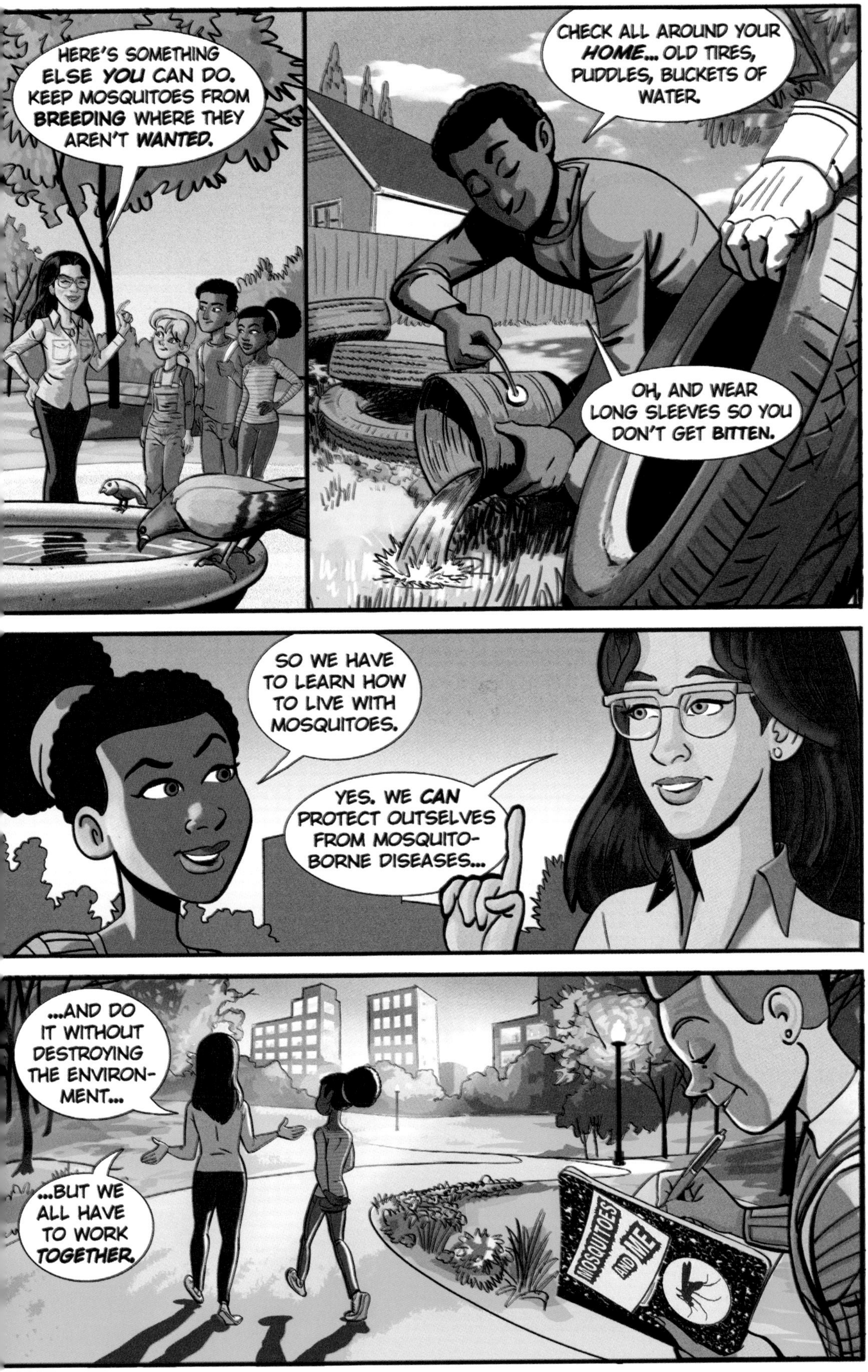

HERE'S SOMETHING ELSE YOU CAN DO. KEEP MOSQUITOES FROM BREEDING WHERE THEY AREN'T WANTED.
CHECK ALL AROUND YOUR HOME... OLD TIRES, PUDDLES, BUCKETS OF WATER.
OH, AND WEAR LONG SLEEVES SO YOU DON'T GET BITTEN.
SO WE HAVE TO LEARN HOW TO LIVE WITH MOSQUITOES.
YES. WE CAN PROTECT OUTSELVES FROM MOSQUITO-BORNE DISEASES...
...AND DO IT WITHOUT DESTROYING THE ENVIRON-MENT...
...BUT WE ALL HAVE TO WORK TOGETHER.
MOSQUITOES AND ME

MOSQUITOES SUCK!

We've all felt it. The twinge. And then it starts. You see the redness. You feel the itch. And there it is—the bump. The notorious mosquito bite. But what really is it?

Most people look at that red itchy bump and call it a mosquito "bite." While other animals that bite, like a shark or a dog, or even a beetle or ant, have big moving jaws, mosquitoes don't. A mosquito mouth is made up of needles that pierce the skin, and a long feeding tube. So, mosquitoes don't bite. They SUCK!

But why do they suck? They suck nectar out of plants for energy. They suck blood out of animals for protein. They do these things to survive. Just as you need food to fuel your body, mosquitoes need sugar-rich nectar to fuel their bodies. Like butterflies and moths, they use their long mouths to sip nectar from flowering plants. This sugar gives them energy to walk, fly, mate, and find blood.

Not all mosquitoes suck blood. Male mosquitoes just want nectar. And while all female mosquitoes want nectar too, it's mosquito moms who suck blood. They do this to nourish their eggs. And they're really sneaky about it! Even before you may feel a twinge, these mosquito moms are probing inside your skin with their mouthparts to locate a small blood vessel.

When she lands on you, a mosquito mom performs a fast and delicate medical procedure with her proboscis (pronounced pro-bah-sis NOT pro-bah-skiss!), the scientific name for her mouthparts. Inside the proboscis are six different parts that pierce your skin and slither under it to find a blood vessel. Some of the mouthparts fold together to create a straw-like tube, and then muscular pumps suck the blood. As she is sucking your blood up, she's simultaneously spitting saliva out! Mosquito spit thins the blood and keeps it flowing—"fast food" for the mosquito!

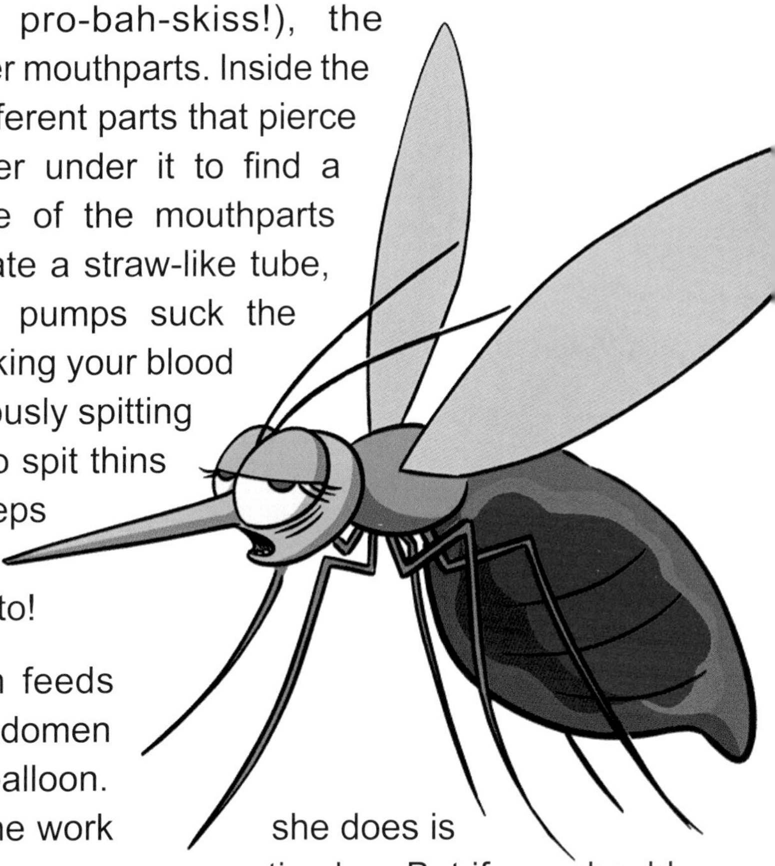

The mosquito mom feeds like this until her abdomen swells up like a balloon. She is so tiny and the work she does is so precise, you may never notice her. But if you should feel her feeding and smack at her in a rage, you'll burst that swollen belly and see a SPLAT of blood.

If you're sensitive to mosquito bites, your body will know that she's been there spitting into your skin. The body's defense system recognizes and responds to the foreign mosquito spit.

Within minutes you may find that itchy red bump at the site of the mosquito "suck."

If she's lucky enough to get away, this mosquito mom will need a big rest to digest all she's eaten—just like you do after a big meal. While she rests, her system digests the enormous blood meal—quickly separating out the water and then sending proteins to her ovaries to generate a big batch of eggs.

After three to five days, the eggs have gone from tiny transparent spheres to chalky-white, plump eggs ready to pop out the end of her abdomen. She flies around searching for just the right spot for her babies to hatch and grow. Some prefer sunny or shady puddles, flooded woodland tree holes, or stinky drains and gutters. Others are happy with the pools of water collected in human-made containers such as garbage cans, plant pots, trash, and children's toys.

Mosquitoes can lay hundreds of eggs at a time. After they deposit their eggs, mosquito moms are hungry again and go on a new hunt for sugar and blood. Meanwhile, larvae hatch from the eggs in the water and look like tiny wiggling worms. In summer these larvae will take about seven to ten days to eat and go through four stages of growth. Then they enter the next stage—called pupae—and in two to three days develop all the organs and structures of an adult mosquito. At this point, they leave the water and become the next generation of mosquitoes that SUCK!

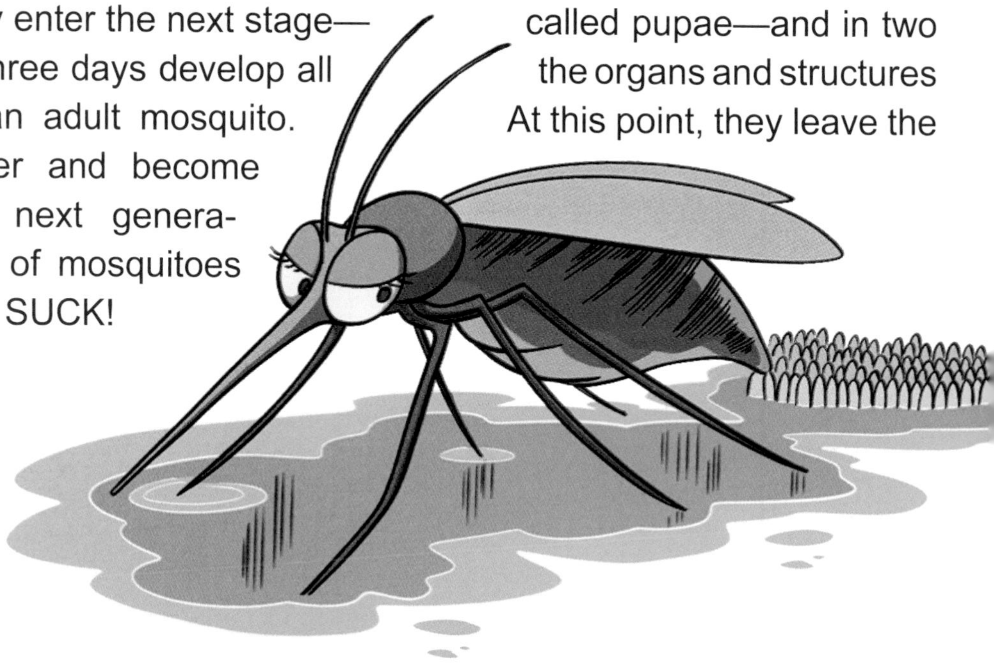

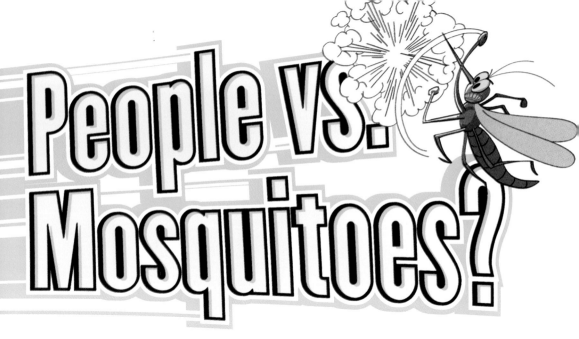

People vs. Mosquitoes?

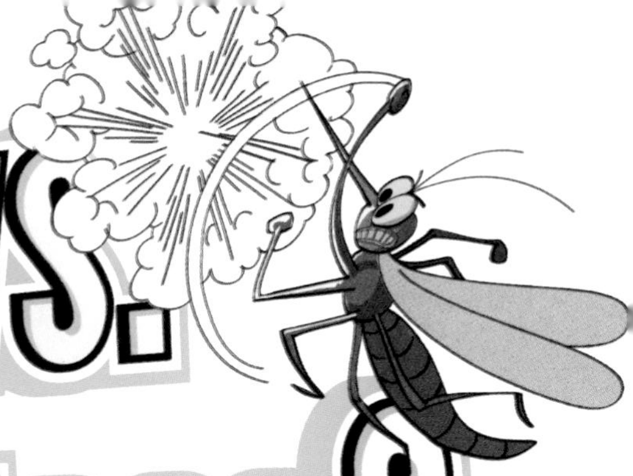

There are good reasons not to like mosquitoes. Some people think about half of the humans who've ever lived have died from diseases spread by mosquitoes. That's a lot of harm to people.

Mosquitoes spread disease and do their damage by sucking the blood of animals, including humans. If an animal is carrying parasites and viruses in its blood, a mosquito could accidentally suck those up. If she's the right kind of mosquito, the parasites and viruses will move through her body and go on to infect the next animal she feeds on. Of all the parasites and viruses we know about, only some are spread by mosquitoes. But these few, called mosquito-borne diseases, kill millions of people each year.

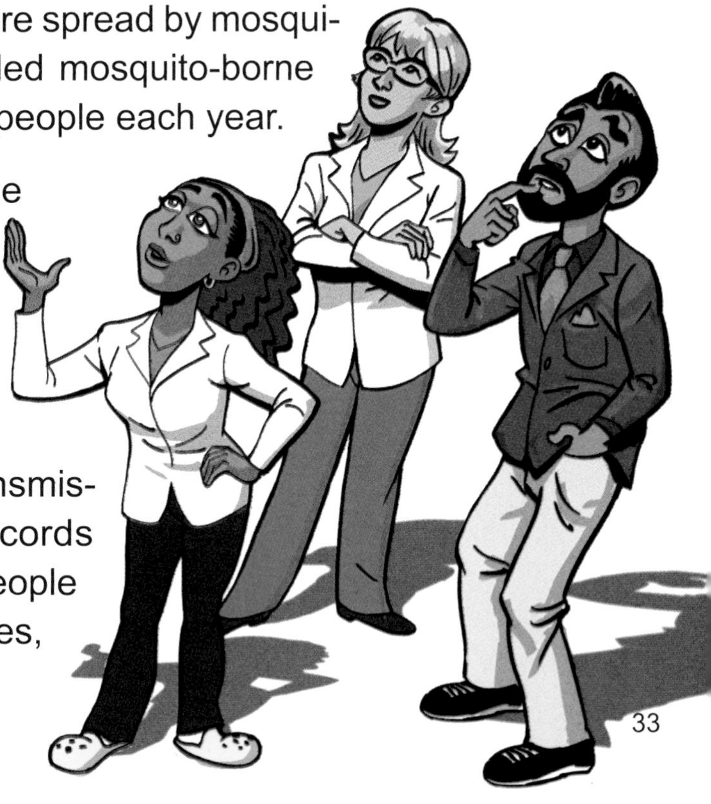

Although mosquito-borne diseases have been around for a really, really long time, scientists only now understand how mosquitoes play a role in disease transmission. As far back as records go, there are tales of people suffering from body aches,

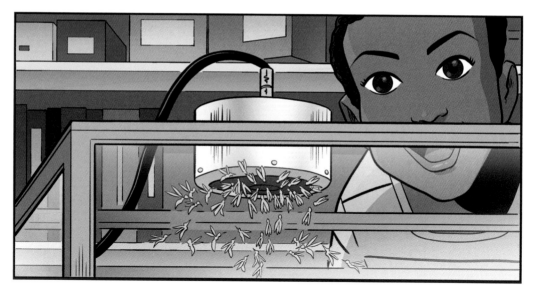

high fevers, weakness, and shaking chills. A sick person would recover for a day or two before the high fever and body chills returned, and then the pattern of symptoms would repeat again.

People wrote about this mysterious illness for centuries. They suspected a link between the symptoms and exposure to marshes and swamps. Some stories said that walking through a marsh brought evil spirits. Others told of *mal aire*—bad air—around a swamp. This is how the disease malaria got its name.

During the time of the United States' Civil War, people didn't understand why Northern soldiers got sick and often died when they ventured into the South. They knew it had something to do with the hot, humid, and swampy landscape. To avoid getting malaria, the soldiers borrowed a remedy long known to some native peoples. They drank quinine, which comes from the bark of a tree. But they didn't know what caused malaria and what it had to do with the soggy, Southern climate.

It wasn't until the late 1800s that scientists connected the dots between the devastating illness in humans and their exposure to swampy landscapes. They did this by conducting lots of experiments: They performed meticulous surgery on mosquitoes using miniature scalpels. They used tiny tweezers to pull out mosquito guts and observe them with microscopes. A scientist noticed that some mosquitoes had small dots that looked out of place in their guts. They also noticed that in these mosquitoes the salivary glands were bursting with eyelash-shaped wiggling objects. Salivary glands make mosquito spit. So this spit and the little wiggling objects are what a mosquito mom leaves behind when she sucks blood.

Other scientists made the connection between parasites in blood, mosquitoes, and the pattern of symptoms they observed in humans. Mosquitoes can get infected by feeding on a sick person and transmit the infection to someone else. It wasn't bad luck or bad air that was causing malaria. It was a parasite passed on by mosquito moms.

Malaria isn't the only mosquito-borne disease we know about. Yellow fever, Zika, dengue, chikungunya, and elephantiasis can also threaten the health of humans. About a million people die each year from the bites of infected mosquitoes! All over the world there are labs where scientists grow millions of mosquitoes to study. They want to learn how to outsmart them. Mosquitoes and their parasites are tricky, but humans are clever, and we have the tools of science on our side. These include the power of our big questions that keep us discovering new things about why and how these tiny insects have such an enormous impact.

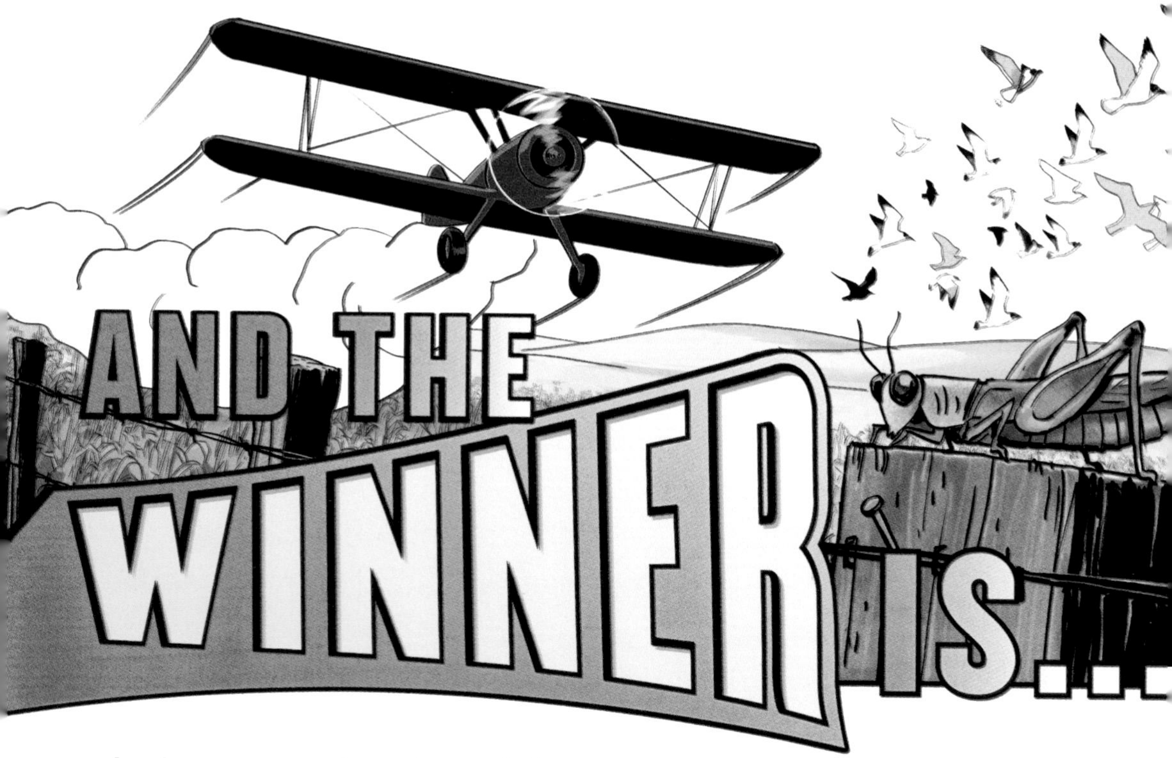

AND THE WINNER IS....

As long as we've shared the planet, mosquitoes and humans have been in a battle of buzzing and biting and swatting and slapping. Whenever mosquitoes seem to be winning, humans have been able to develop a defense until there's a new mosquito challenge. This raises a big question: If there were an ultimate battle for world domination, who would win—mosquitoes or humans?

On the surface, the answer seems simple: humans will win the battle. We are much bigger, and with our big brains we can invent tools for the fight. In our early history, we built fires and used plants to repel mosquitoes. We even used other animals who feed on mosquitoes, like bats and barn swallows, to defend us. We cleared swamps and wetlands to get rid of mosquito breeding grounds. And, where possible, we put screens over windows or nets over our beds to keep mosquitoes away.

Later, we developed chemicals to repel or kill mosquitoes. Some of these chemicals we had to create and others existed naturally in plants. The synthetic chemical DEET that is found in most bug sprays was created by scientists. There are also natural

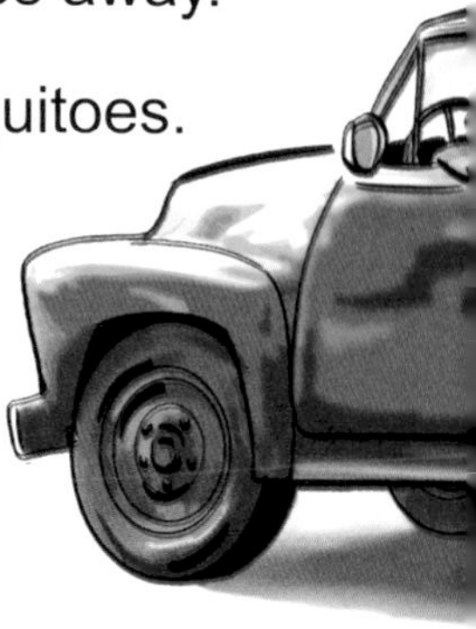

chemicals used for insect control that scientists discovered by studying plants. And now scientists manipulate mosquito DNA so they can't transmit parasites and viruses. So people are out-smarting mosquitoes, right?

When we look deeper, the answer gets less clear. Mosquitoes may be small, but they are mighty, and humans are easily led astray. People around the world once thought that DDT, a synthetic chemical, was the answer to protecting everyone from mosquitoes and mosquito-borne diseases. Trucks sprayed neighborhoods with DDT fog. After contact with this deadly poison, most—but not all—mosquitoes died. This meant fewer mosquito bites! This success inspired more people to use DDT in battle against other insects, such as moths and beetles, that destroyed trees, food crops, and cotton.

Unfortunately, we didn't know that DDT remains in the environment for a long time. Other animals were harmed when DDT got into the water and into microscopic aquatic animals. These tiny animals were eaten by mosquito larvae that were then eaten by little fish who were eaten by bigger fish that were finally eaten by birds. In this way, DDT became concentrated as it moved from animal to animal in a food web. Eventually, with the help of environmental advocates,

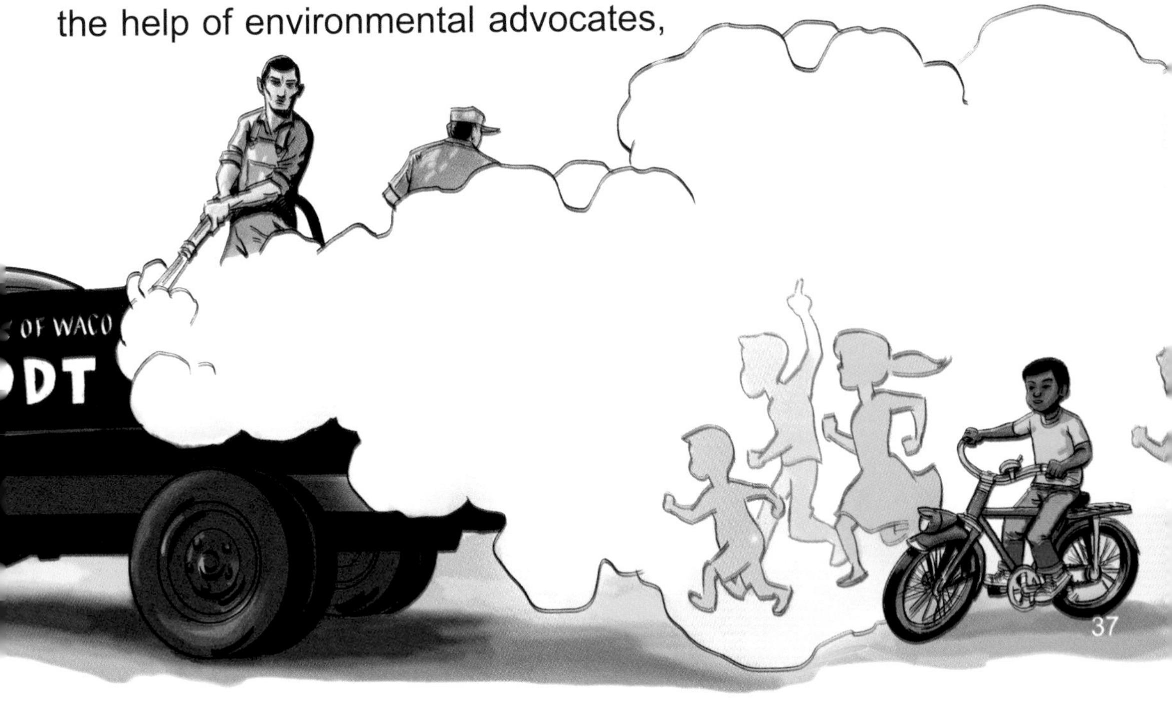

OF WACO
DT

politicians, and scientists, we realized there was 10 million times more DDT in the birds than what had been used for insect control to begin with! The U.S. eventually banned the production and use of DDT.

The human quest to win the battle with mosquitoes has led to surprising results. Remember how some of the mosquitoes survived the DDT fog? Suppose the survivors were a male and a female, and they reproduced to start a new generation of mosquitoes. Most likely, their offspring would also be resistant to the DDT, just like their parents. Each time these mosquitoes pass the trait on to the next generation, there are more resistant mosquitoes. In this way, mosquitoes have been able to survive humans' defensive strategies.

Today, scientists use different strategies to control mosquitoes. A main goal is to minimize impact on other animals and the environment. Ironically, this has meant making more, not fewer, mosquitoes! For example, a type of bacteria called *Wolbachia* can kill some mosquitoes when they are young adults, and it affects how well they can carry disease pathogens. Scientists are purposefully growing *Wolbachia*-infected mosquitoes and releasing them to create new populations that don't transmit deadly viruses. These mosquitoes may still buzz around your ears, suck your blood, and leave itchy red bumps, but their spit is clean—no viruses!

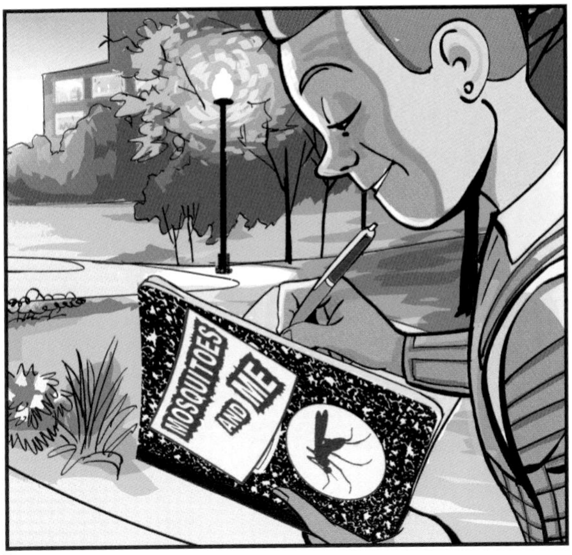
MOSQUITOES
AND
ME

Scientists are also changing mosquito DNA. They are altering the way a malaria parasite grows inside a mosquito so that the parasite dies and can't be transmitted. They are even transforming anatomy so that a mosquito mom's wobbly proboscis can't pierce the skin, or an odor-blind mosquito can't find human blood to suck.

But if we've learned one thing from our battle with mosquitoes, it is that mosquitoes will always adapt. Throughout history, they've been a formidable opponent: small, sneaky, numerous, and persistent.

So is all lost? Humans have a powerful weapon—the ability to think critically—and this will help us outwit mosquitoes while protecting the balance of life. With coexistence as the goal, the new challenge will be to talk not about me *OR* mosquitoes but mosquitoes *AND* me.

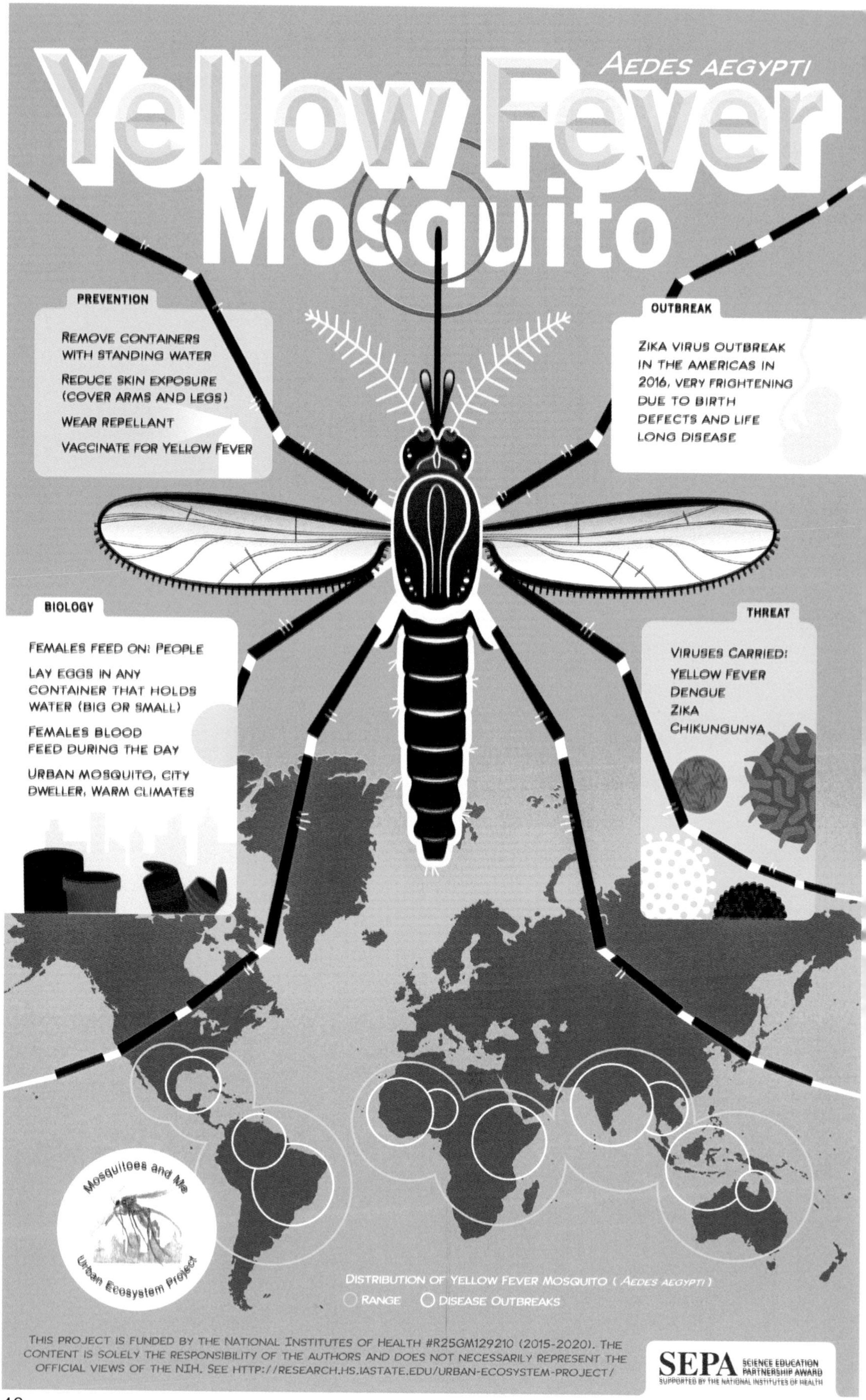

Yellow Fever Mosquito
AEDES AEGYPTI
PREVENTION
REMOVE CONTAINERS WITH STANDING WATER
REDUCE SKIN EXPOSURE (COVER ARMS AND LEGS)
WEAR REPELLANT
VACCINATE FOR YELLOW FEVER
OUTBREAK
ZIKA VIRUS OUTBREAK IN THE AMERICAS IN 2016, VERY FRIGHTENING DUE TO BIRTH DEFECTS AND LIFE LONG DISEASE
BIOLOGY
FEMALES FEED ON: PEOPLE
LAY EGGS IN ANY CONTAINER THAT HOLDS WATER (BIG OR SMALL)
FEMALES BLOOD FEED DURING THE DAY
URBAN MOSQUITO, CITY DWELLER, WARM CLIMATES
THREAT
VIRUSES CARRIED:
YELLOW FEVER
DENGUE
ZIKA
CHIKUNGUNYA
Mosquitoes and Me
Urban Ecosystem Project
DISTRIBUTION OF YELLOW FEVER MOSQUITO (AEDES AEGYPTI)
RANGE
DISEASE OUTBREAKS
THIS PROJECT IS FUNDED BY THE NATIONAL INSTITUTES OF HEALTH #R25GM129210 (2015-2020). THE CONTENT IS SOLELY THE RESPONSIBILITY OF THE AUTHORS AND DOES NOT NECESSARILY REPRESENT THE OFFICIAL VIEWS OF THE NIH. SEE HTTP://RESEARCH.HS.IASTATE.EDU/URBAN-ECOSYSTEM-PROJECT/
SEPA SCIENCE EDUCATION PARTNERSHIP AWARD
SUPPORTED BY THE NATIONAL INSTITUTES OF HEALTH

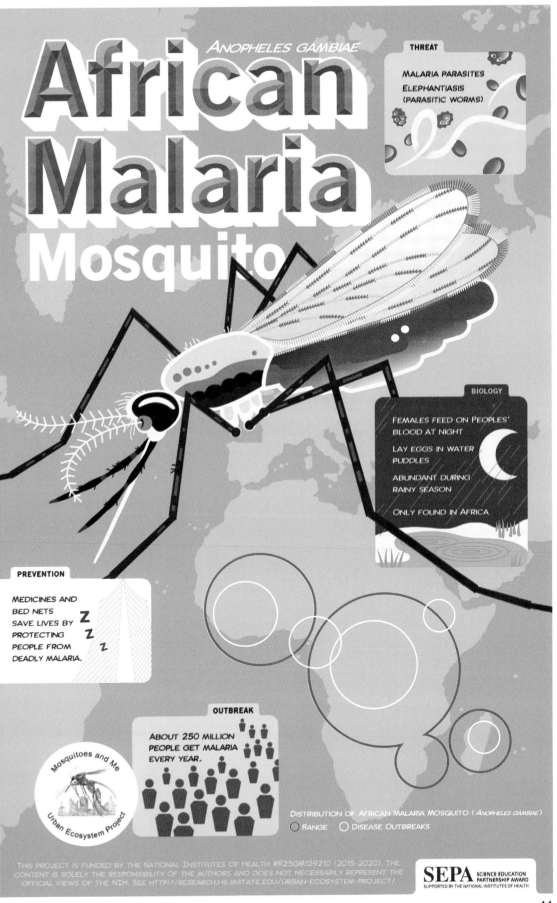

African Malaria Mosquito

Anopheles gambiae

Mosquito

THREAT

MALARIA PARASITES

ELEPHANTIASIS (PARASITIC WORMS)

BIOLOGY

FEMALES FEED ON PEOPLES' BLOOD AT NIGHT

LAY EGGS IN WATER PUDDLES

ABUNDANT DURING RAINY SEASON

ONLY FOUND IN AFRICA

PREVENTION

MEDICINES AND BED NETS SAVE LIVES BY PROTECTING PEOPLE FROM DEADLY MALARIA.

OUTBREAK

ABOUT 250 MILLION PEOPLE GET MALARIA EVERY YEAR.

Mosquitoes and Me

Urban Ecosystem Project

DISTRIBUTION OF AFRICAN MALARIA MOSQUITO (*ANOPHELES GAMBIAE*)

○ RANGE   ○ DISEASE OUTBREAKS

THIS PROJECT IS FUNDED BY THE NATIONAL INSTITUTES OF HEALTH #R25GM129210 (2015-2020). THE CONTENT IS SOLELY THE RESPONSIBILITY OF THE AUTHORS AND DOES NOT NECESSARILY REPRESENT THE OFFICIAL VIEWS OF THE NIH. SEE HTTP://RESEARCH.HS.IASTATE.EDU/URBAN-ECOSYSTEM-PROJECT/

SEPA SCIENCE EDUCATION PARTNERSHIP AWARD
SUPPORTED BY THE NATIONAL INSTITUTES OF HEALTH

41

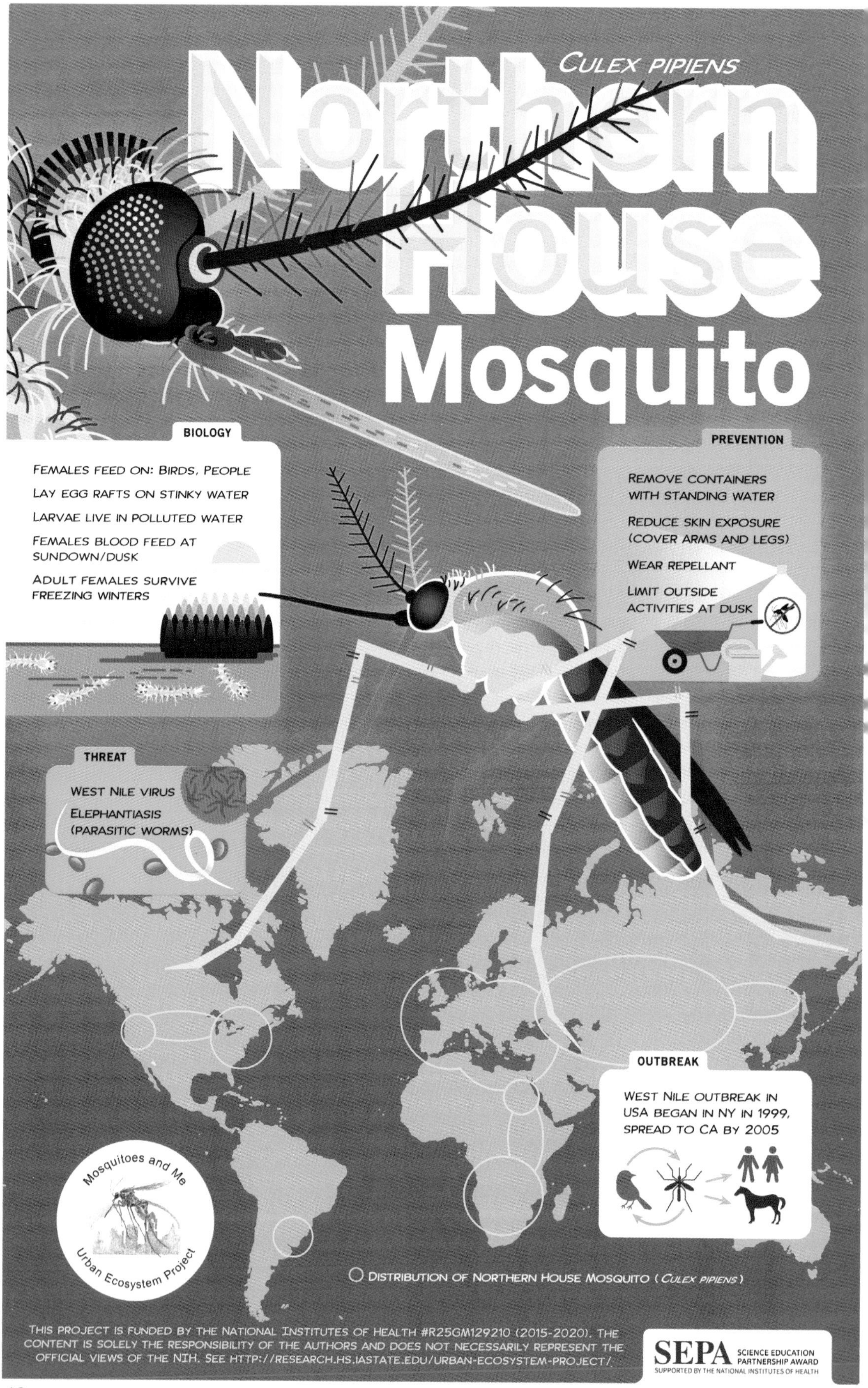

Culex pipiens
Northern House Mosquito
BIOLOGY
Females feed on: Birds, People
Lay egg rafts on stinky water
Larvae live in polluted water
Females blood feed at sundown/dusk
Adult females survive freezing winters
PREVENTION
Remove containers with standing water
Reduce skin exposure (cover arms and legs)
Wear repellant
Limit outside activities at dusk
THREAT
West Nile virus
Elephantiasis (parasitic worms)
OUTBREAK
West Nile outbreak in USA began in NY in 1999, spread to CA by 2005
Mosquitoes and Me
Urban Ecosystem Project
Distribution of Northern House Mosquito (Culex pipiens)
This project is funded by the National Institutes of Health #R25GM129210 (2015-2020). The content is solely the responsibility of the authors and does not necessarily represent the official views of the NIH. See http://research.hs.iastate.edu/urban-ecosystem-project/.
SEPA Science Education Partnership Award
Supported by the National Institutes of Health

THIS PROJECT IS FUNDED BY THE NATIONAL INSTITUTES OF HEALTH #R25GM129210 (2015-2020). THE CONTENT IS SOLELY THE RESPONSIBILITY OF THE AUTHORS AND DOES NOT NECESSARILY REPRESENT THE OFFICIAL VIEWS OF THE NIH. SEE HTTP://RESEARCH.HS.IASTATE.EDU/URBAN-ECOSYSTEM-PROJECT/

43

ABOUT THE AUTHORS AND ARTISTS

Lyric Bartholomay is a professor of pathobiological sciences at the University of Wisconsin-Madison and a director of the Midwest Center of Excellence in Vector Borne Disease. With training in microbiology and medical entomology, she investigates the transmission of pathogens by arthropod vectors through laboratory and field studies. She wants more young people to study entomology, and blood suckers in particular!

Bob Camp is an American animation director, cartoonist, and writer with forty years' experience in comics and animation for TV and film. Bob is best known for his work on Nickelodeon's *The Ren & Stimpy Show.* He has taught animation at SVA in New York for eight years and runs Boblab Studios in New Jersey.

Michael Cavallaro has worked in comics and animation since the early 1990s. His comics include *Parade*, *The Life and Times of 28, Impossible Incorporated*, *Foiled*, *Decelerate Blue*, and the Nico Bravo series. www.mikecavallaro.com

Judy Diamond is curator and professor at the University of Nebraska State Museum. She is the lead author of the comics *World of Viruses* (University of Nebraska Press, 2012) and *Occupied by Microbes* (Zea Press, 2019).

Mike Edholm is an award-winning graphic designer, illustrator, writer, and cartoonist who devotes his time exclusively to freelance cartoon and illustration work and the promotion of cartoons, humorous art, and illustration. www.mikedholm.com

Sara Erickson is a medical entomologist who studies the interactions between mosquitoes and the pathogens they transmit. She has worked in mosquito laboratories all around the world and is passionate about science education.

Bob Hall is an artist and writer who worked for Marvel Comics, drawing most of their major characters including the Avengers, Thor, Spider-Man, the Fantastic Four, Captain America, the Black Widow, and Doctor Doom. His work includes *Batman*, *Shadowman*, and *Carnival of Contagion.*

Katherine Richardson Bruna is a professor of sociocultural studies at Iowa State University. Trained as an educational anthropologist, she uses the tools of observation to examine and shape the cultures that arise in particular learning contexts. Her work explores dynamics of language, identity, and power in science education and how informal learning experiences can interact with these to enhance science agency for historically excluded students. Dr. Richardson Bruna remembers buying her first comic book on a family trip to the Oregon coast and being pursued by mosquito swarms on her grandparents' farm in Minnesota. Never in her wildest dreams did she think her professional life would include either.

Aaron Sutherlen's work is grounded in the principles of design authorship and social responsibility. His work has included designing exhibitions and installations for a number of nonprofit organizations focused on creating positive social change for the community.